U0396387

中庸建构

浙江大学建筑系高裕江工作室建筑创作 2004—2014

APPROPRIATE CONSTRUCTION

东南大学出版社·南京

序 言

20 年前的秋天，本科毕业十年的学生高裕江手中拿着一叠设计作品资料来到东南大学建筑研究所拜师求学的情景犹在眼前。当时的青年建筑师高裕江已取得不俗的设计业绩：作为主创建筑师，1987—1991 年他设计创作的第一项作品——绍兴鲁迅电影院工程，荣获浙江省优秀设计二等奖、建设部优秀设计三等奖，该工程是在华东地区主要设计单位参与的竞赛中获得第一名而实施的公共建筑；其主创的第二项作品，文化部第一家戏剧类专业博物馆——越剧博物馆工程，也获得浙江省优秀设计三等奖。此两项作品均入选浙江省首届建筑设计创作奖（全省共 10 项），小小年纪居然有其二，这样的初创业绩确实不简单了。

其在杭州市建筑设计研究院师从程泰宁院士期间，从程先生那里学到了丰富的专业知识和可贵的创作经验，在协助程先生创作加纳国家剧院工程、马里议会大厦工程及河姆渡遗址博物馆工程、北京民族大厦等设计及竞赛过程中，均受益匪浅。其因设计业绩突出，在 1995 年破格获得了高级职称资格。

当年这些成绩的取得，除了其勤奋和天赋外，同其在校时所打下的优秀扎实的学业基础分不开。早在东大本科期间，其优秀的专业素养已展露出来：其一，他的美术老师、画家梁蕴才先生就十分欣赏他的水彩写生作品。有一次在南京栖霞寺内写生，当高裕江的水彩写生作成时，为师的梁先生非常激动，叫小组的二十来位学生均来观摩。在后来的东大与日本爱知工业大学的学生优秀作品交流中，选其多幅水彩实习作业去日本交流巡展。其二，当年的东南大学建筑学研究生入学考试，在全国 150 多名考生中，其快题设计成绩成为两名高分中的一位，说明了其所具有的优秀设计专业基础。

今天，《中庸建构——浙江大学建筑系高裕江工作室建筑创作（2004—2014）》的书稿展现在我面前，这是一份辛勤耕作的设计成果。作者说，他不够勤奋和努力，显然这是他的谦虚表现。书中二十八个项目，类型广泛，构思独特，反映其作为主创建筑师深厚的设计功力。很多作品是其参加国际、国内竞标获得第一名的方案及其实施的项目，例如江苏省图书馆新馆工程（2001—2007），是在国内外几十个竞标方案中最终中标实施的项目，我参访过建成后的作品，其功能整构的合理性、内部空间的丰富性，以及空间形态与建筑形式风格的融汇手法，自然地显示出其较高的创作水准，其成果得到了省文化厅和业主的称赞以及社会的广泛认同；再如，其受邀参加的天津图书馆国际竞标方案，同其他三位来自国际著名设计大师竞争，其方案得到了当时作为评委之一的彭一刚院士的高度认同和赞誉，虽然因各种原因未能中标，但其创作水平还是上升到了一定的高度；再如其主创的浙江大学舟山校区（海洋学院）的规划建筑方案，在极其不规则的山地、湖间用地内，巧妙地保留了场地中的部分山体，并合理运用“教育轴”和“组团式”相整构的手法，较自然地建立起丰富而有序的校园空间环境，其实施的规划及建筑概念方案得到了浙江大学及舟山市政府的高度认可。尽管后期因故设计变更，校园建设在基本保持原规划及空间形态的基础上，单体建筑屋顶采用浙江大学玉泉老校区大屋顶的形式，其因持不同意见而放弃了后期设计主持，此固然有点可惜，但也展示出其作为学术型建筑师所具有的理性和执着的职业素质。

我的学生刚进入创作的黄金时期，真诚地祝福和希望，其在今后的建筑创作生涯中，继续努力，为社会奉献更多优秀的设计作品。

中国科学院院士

国家设计大师

PREFACE

In the fall 20 years ago, a student Gao Yujiang, who had graduated for ten years,holding a bundle of design work data, came to the architectural institute of southeast university for deep study. The scene is still in front. At the time, the successful young architect Gao Yujiang had achieved good design performance. As a principal architect in 1987—1991, he designed his first work ,Lu Xun Theater in Shaoxing, which won the second prize of excellent design of Zhejiang Province and third prize of outstanding design of Ministry of Construction . The project won the first prize of the competition participated by main design units in East China and then was put into practice. The second creative work by Gao was Shaoxing opera museum project, the first professional drama museum of Ministry of Culture, which also won the third prize for outstanding design in Zhejiang Province. And the two works were both selected in the first architectural design and creation awards of Zhejiang Province(10 items in total in Zhejiang Province). At an early age, Gao had been excellent and the start-up performance was excellent.

During the practicing in architectural design and research institute in Hangzhou , Gao learned rich professional knowledge and precious experience from Cheng Taining , benefiting a lot by assisting Cheng in the Ghana national theatre project, Mali parliament building engineering , Hemudu ruins museum and Beijing national building design. Because of his outstanding design performance, Gao won the senior title of professional qualifications exceptionally in 1995.

Besides his hard work and talent, Gao obtained the achievements with his fine and solid academic foundation built in school. As early as during the period of undergraduate course, his speciality accomplishment had shown. His art teacher, the painter Liang Yuncai appreciated his watercolor painting works so much. Once students painted in Qixia temple of Nanjing , as Gao's watercolor painting was finished, Mr Liang was very excited, and let me group of over twenty students watch it. In the exchange of students' outstanding works between Southeast University and Aichi Institute of Technology,Gao's several watercolor paintings were chosen to the tour in Japan. In the postgraduate entrance examination of architecture in the southeast university that year, Gao's fast design work became one of the two highest marks in more than 150 candidates ,which illustrated his excellent design professional basis.

Today, the manuscript of *Appropriate Construction: Architectural Creation (2004—2014) by Studio Gao Yujiang from the Department of Architecture,Zhejiang University* is delivered to me, it is a design result of hard work. The author thinks of himself not hardworking, which obviously displays his modesty. The book covers twenty-eight projects, with wide range of types and unique ideas, reflecting his profound design capability as the principal architect . Several projects in the book had won the first prize in domestic and international bidding schemes and then came into practice, such as the new library project of Jiangsu Province (2001—2007) .It was the successful design proposal among dozens of the domestic and international bidding schemes and put into implementation ultimately . I visited the building afterwards, the rationality of the function structure , the abundance of the interior space, and fusion technique dealing with space form and architectural form style , naturally showed his high creation level. The results obtained praise from the province cultural department and the owner,as well as wide acceptance from society. Gao was ever invited to participate in the international bidding scheme of Tianjin Library, with three other international famous design masters. His work obtained highly recognition and praise from academician Peng Yigang who worked as one of the judges at the time. Although Gao didn't win finally because of various reasons, but the level of the work still rose to a certain height. In the planning construction scheme about Zhoushan Campus (Ocean College) of Zhejiang University , Gao cleverly retained part of the mountain in the extremely irregular topography including mountains and lakes, and naturally built a rich and orderly campus space environment with the reasonable use of the technique integrating "education axis" and "group type" . The planning and architectural concept got high approval from the Zhejiang University and Zhoushan city government. Since the design changed afterwards, the individual building roof finally adopted the form of the big roof from Yuquan campus of Zhejiang University, although the campus construction kept the original planning and spatial form. Gao abandoned the later design on account of different opinions, which may be a pity, but showed the professional quality of rationality and persistence as an academic architect.

My student Gao is coming into the golden period of creation, I sincerely wish and hope him to continue to work hard and to contribute to society with more excellent design works in his future architectural career.

Qi Kang Academician

National Design Master

目 录 CONTENTS

南京玄武湖隧道控制中心

The Control Center of Nanjing Xuanwu Lake tunnel

实 施 工 程　2001—2004
获 奖 情 况：国家优秀建筑设计铜奖、全国优秀建筑设计（建设部）二等奖、江苏省优秀建筑设计二等奖
建筑师团队：高裕江、屠子纲、陈波、刘文军
设计主体单位：南京市建筑设计研究院

南京玄武湖隧道控制中心是当时我国最长城市隧道的控制中心。设计上从城市设计层面，整合其与明城墙的环境同构关系；形态及体量上以“平直”“简洁”的现代建筑语汇同古城墙进行有机对话，从而提升历史地段的建筑环境品质。运用“过楼”“上下复合中庭”的空间模式，突显景观视廊及动静空间的形态特色。

石料、合金、玻璃等三材的协同构筑，使其不仅拥有“明城遗风”的神韵之意，而且也以明快、典雅的风格展露其清新不俗之感。

The control center of Nanjing Xuanwu Lake Tunnel is the control center of longest urban tunnel in our country. The project is from an aspect of urban design, reconstructed isomorphically related to the environment of Ming Dynasty City Wall; its morphology and mass dialogue with the ancient city wall using modern architecture terms such as "straight" and "concise", so as to improve the quality of architecture environment in the historical area. Using the spatial model of "Guo building" and "up-down combined atrium", the project highlights the morphologic characters of landscape visual corridor and dynamic-static space.

Stones, metals and glasses are synthesized so as to not only possess the charm of "Ming City Relique", but also unfold the pure and fresh feeling in a vivid and elegant manner.

总平面图 | Site Plan

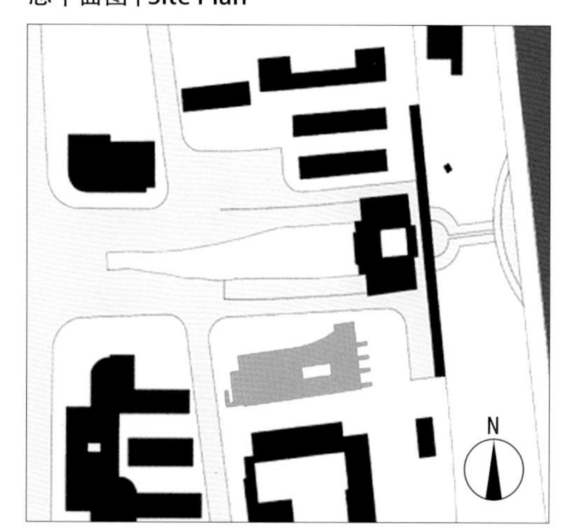

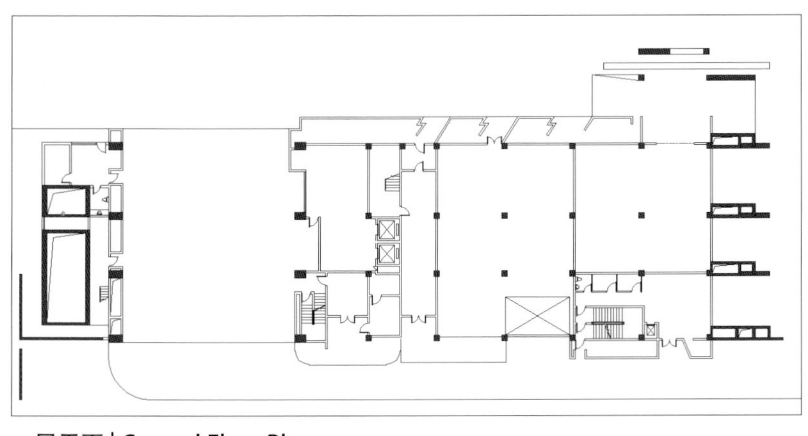

一层平面 | Ground Floor Plan

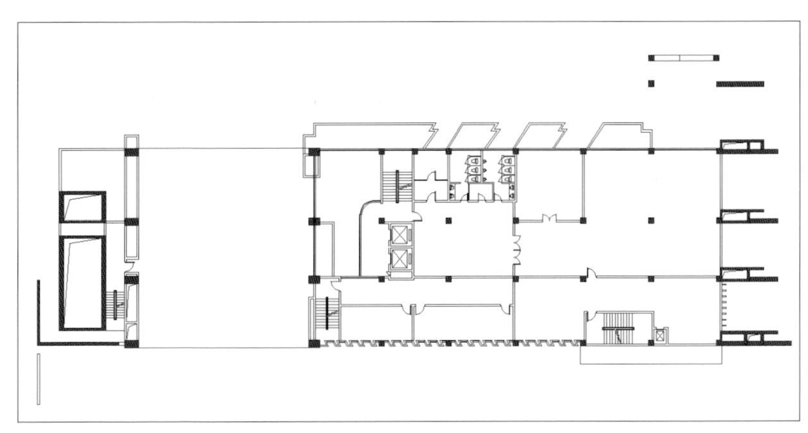

二层平面 | Second Floor Plan

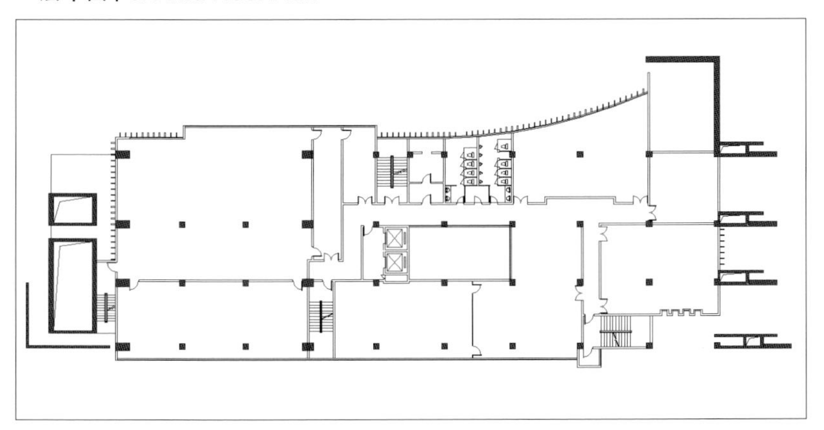

四层平面 | Fourth Floor Plan

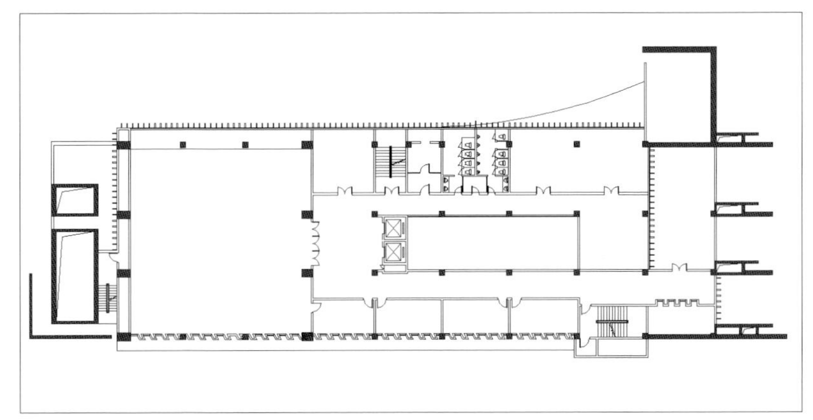

七层平面 | Seventh Floor Plan

南京图书馆

Nanjing Library

实 施 工 程 国际竞标（第一名）/2001—2007
获 奖 情 况：国家优质工程金质奖（鲁班奖）、江苏省优秀设计一等奖
建 筑 师 团 队：高裕江、李永漪、罗明辉、钱晓青、黄龙明、周研、刘文军、王小江等
主体设计单位：南京市建筑设计研究院

南京图书馆即江苏省图书馆，藏书量 1200 万册，阅览位 3000 座。

设计通过“南北廊架”与东西向“Y”形空间的组合，使图书馆交通空间合理有机，并通过廊桥的界定，动静空间得以明晰和强化。而椭圆形空间的介入，不仅使内外空间形态得到整构和融合，东西景观空间随之贯通，进而使城市广场空间与图书馆大堂空间形成有机过渡，凸显空间形态的公民性、丰富性和艺术性。

Nanjing Library, also named as Jiangsu Library, holds twelve million books and three thousand seats.

The combination of "north-south corridor" and east-west Y-shaped space creates rational and organic traffic spaces for the library; moreover, according to the definition of Lang Bridge, the dynamic and static spaces come to be clear and highlighted. The intervention of an elliptic space not only reconstructs and fuses the inside and outside architectural spaces, but also joins up east and west landscape spaces, so as to form an organic transition between the urban entrance square and the main hall of the library, and meanwhile, pinpoints the spatial morphology which is human-oriented, colourful and artistic.

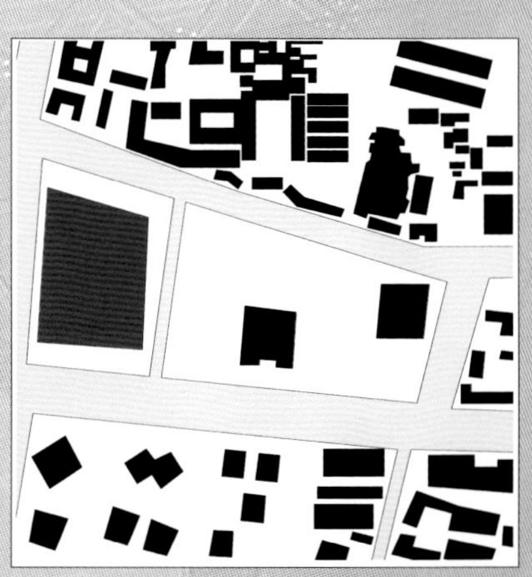

总平面图 | Site Plan

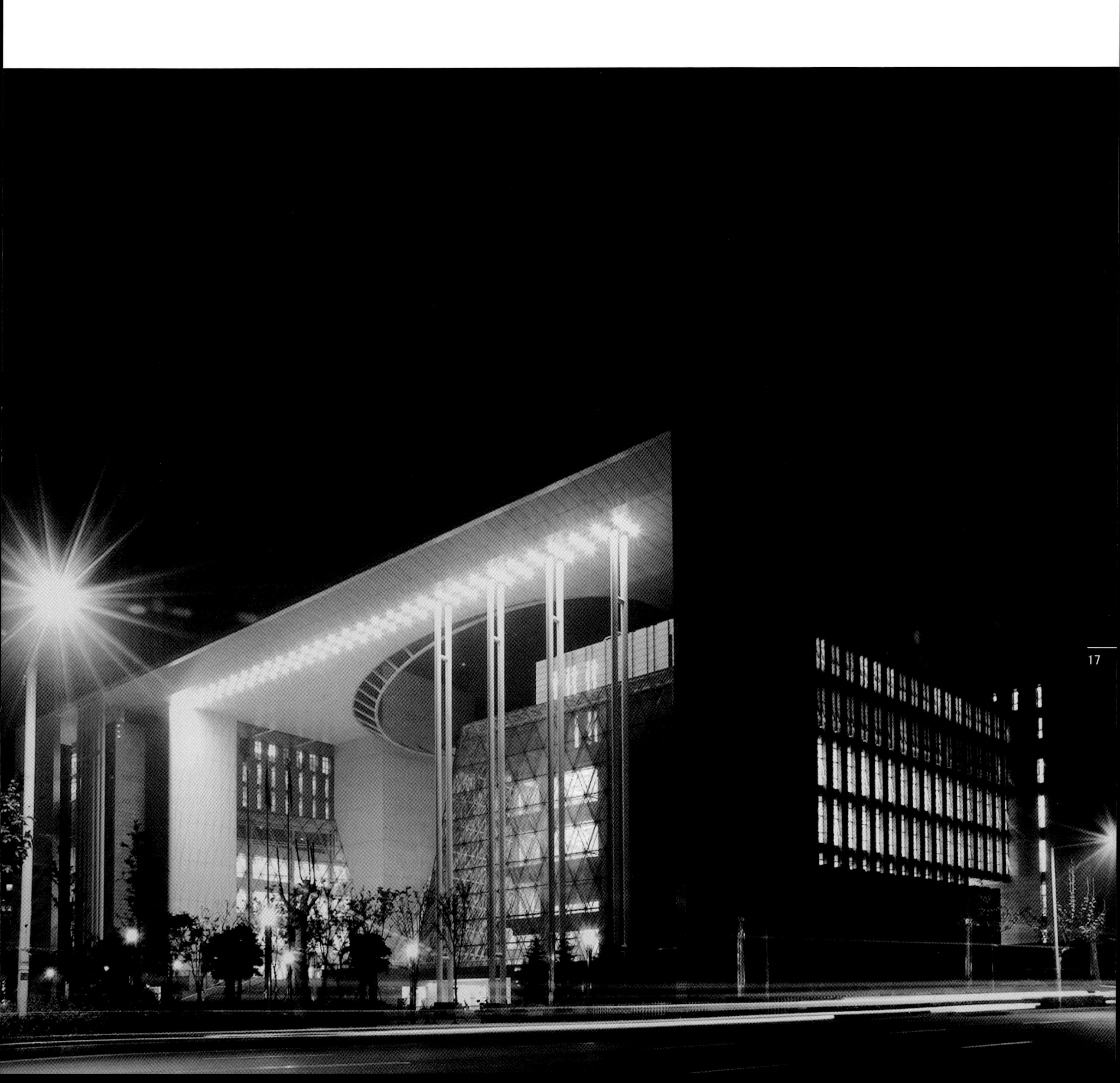

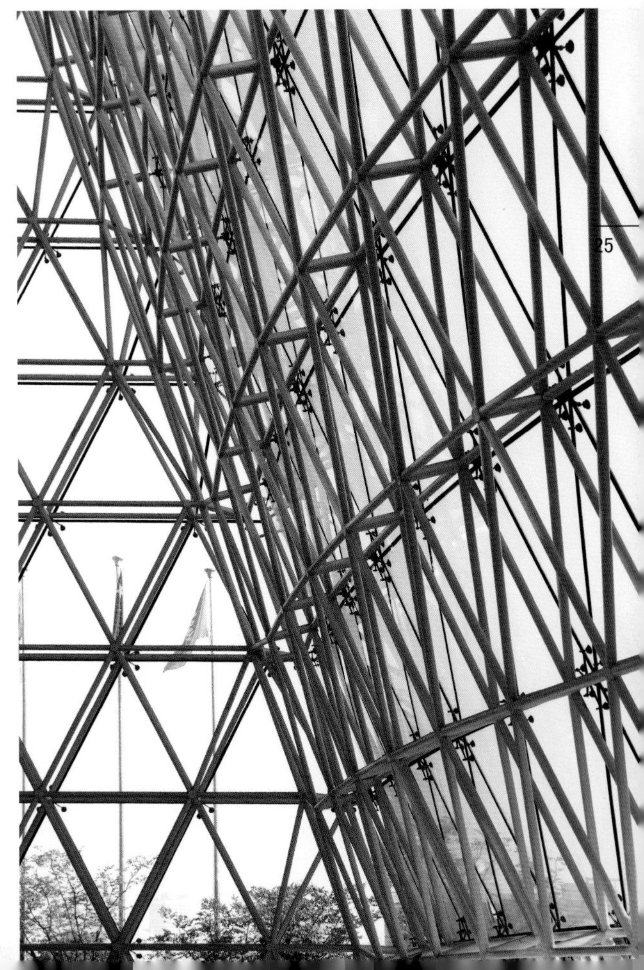

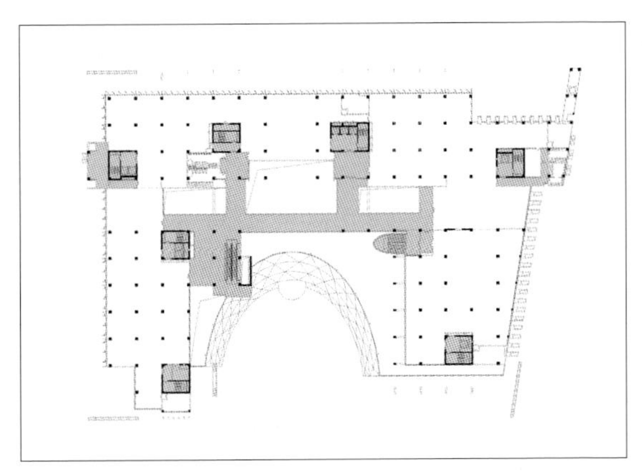

四层平面 | Fourth Floor Plan

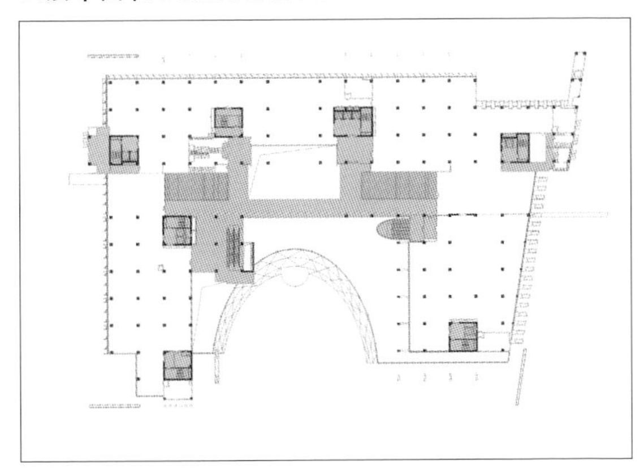

三层平面 | Third Floor Plan

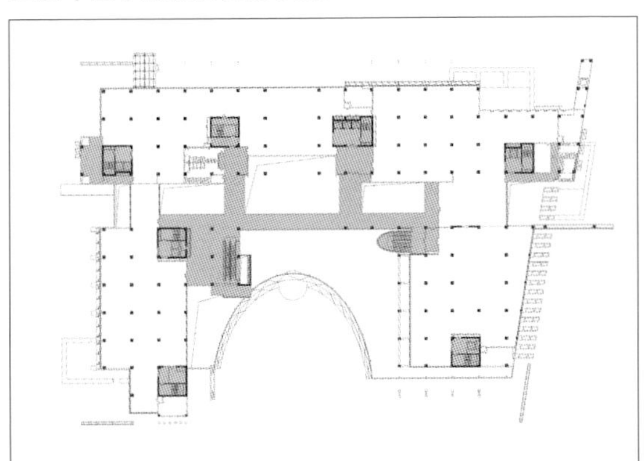

二层平面 | Second Floor Plan

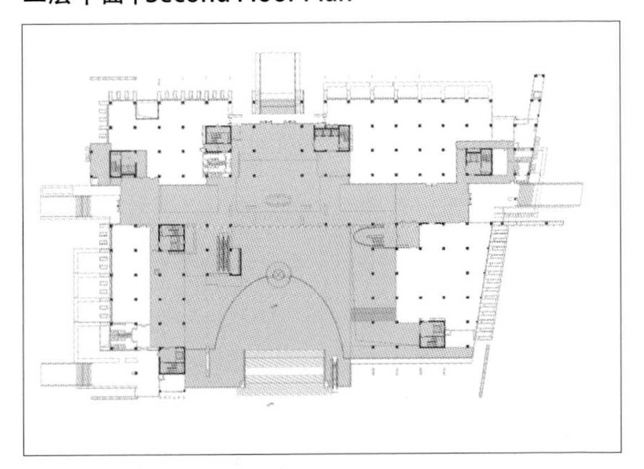

一层平面 | Ground Floor Plan

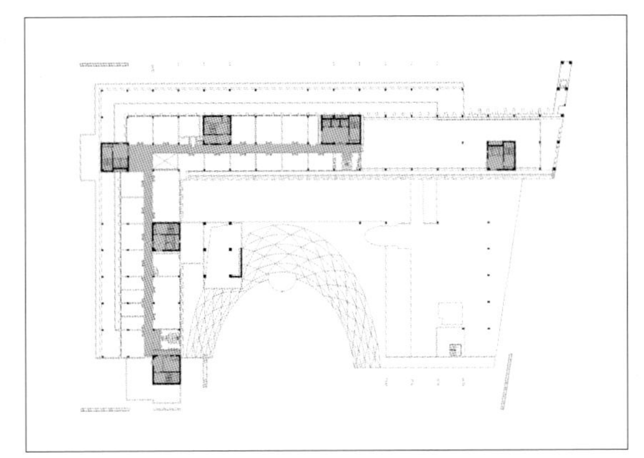

八层平面 | Eighth Floor Plan

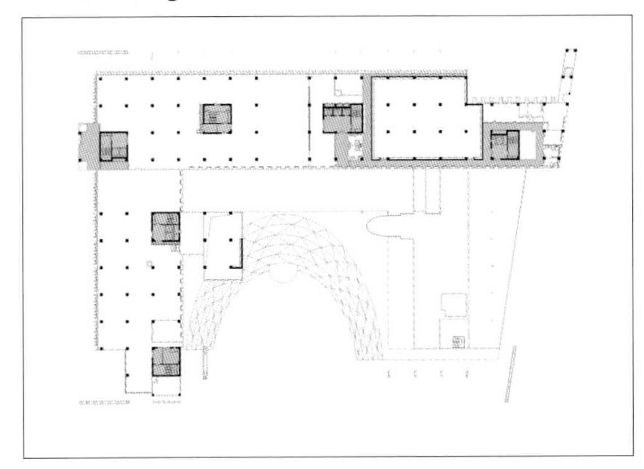

七层平面 | Seventh Floor Plan

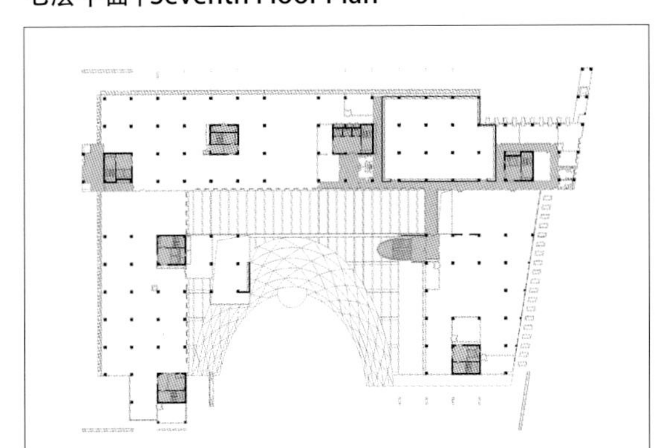

六层平面 | Sixth Floor Plan

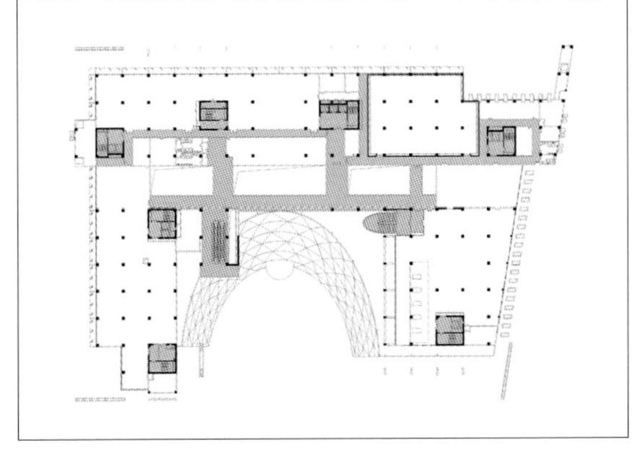

五层平面 | Fifth Floor Plan

NO-004
NO-003

黄山徽州风情小镇

A Planning Scheme for Cultural Town, Huizhou, Huangshan

概 念 规 划　2005
建筑师团队：高裕江、范旭明、章文杰、
　　　　　　杜书凯、曹晓敏、夏琴

在古城之边，一个相对完整的空间里，打造一个既属于历史，又属于现在，更属于未来的产品，这应该是什么？应该是一个无愧于历史的生活空间。在这个空间里，你能感悟到历史，也消费未来。这是一个用二十一世纪初的视角来诠释徽州的一个精品，在这里，用传统的营造理念和手法，用传统的石头、木料、砖头构成一个与历史安静相依的小镇，它也是古城的一部分，地理上它们完美相连，形态上它们互相协调，功能上它们互相补充，它们是今天营造的文化，也是留给后人的文物。

At the edge of the ancient town, in a relatively complete space, to manufacture a product belonging to the history, the present and even the future, what kind of thing it should be? It should be a living space which lives up to history. In this space, you can not only realize the history, but also consume the future. It should be a masterpiece to represent Huizhou. Here, depending on the traditional design methodology, a small town formed by localized stones, wood and bricks is connected with history. It should also be a part of the ancient town. The ancient and the newly-built town, geographically, connect well with each other; morphologically, they are harmonious with each other; functionally, they reinforce each other. They are not only present culture, but also cultural relics.

总平面图 | Site Plan

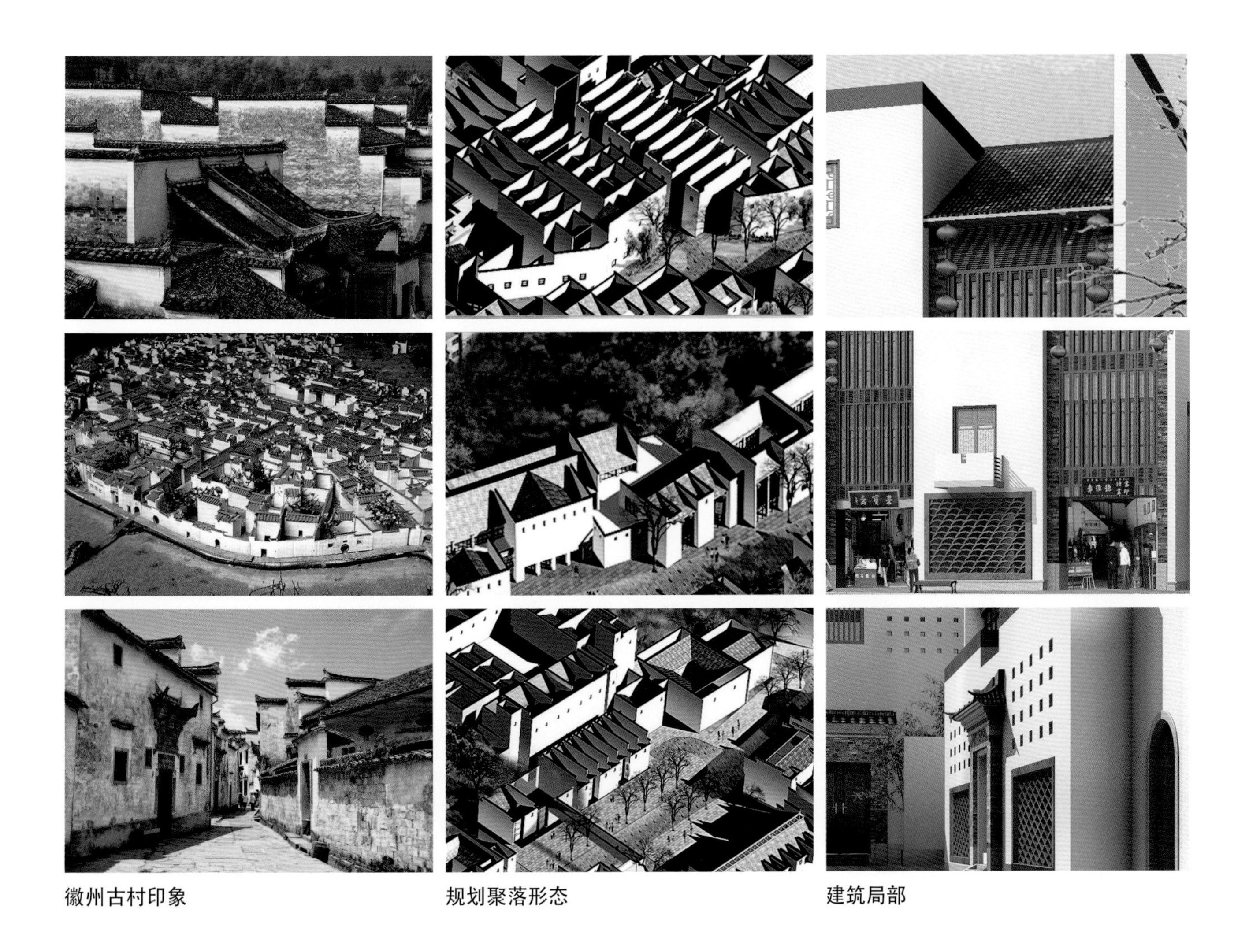
徽州古村印象　　规划聚落形态　　建筑局部

宁波博物馆

Ningbo Museum Design

设 计 方 案　国际竞标 /2005
建筑师团队：高裕江，章文杰，李杰勇，斯春霖，夏琴，曹晓敏

展示与保护其具有七千年文明史和一千两百年州城史的宁波博物馆设计理念中，通过“内外之海”的设定以隐喻宁波的海域地理特征。建筑形象源自古船的“有相”，变异成类船形的建筑形体的“无相”，将成为一种文化形态。故而，“有相”与“无相”的建筑形式，也自然体现了“有形”或“无形”的文化价值。在空间建构中，取之“内海”之端的下沉入口空间，具有了一种空间和时间上的张力，一种历史的深度与时空的跨度。空间环境的光与影、虚与实、动与静的对比，折射出其儒雅、和美的品质。

According to the design concept of presentation and protection of seven thousand years' history of civilization, and one thousand and two hundred years' history to be a city or town, the design basing on the hypothetic model of "inland and outside sea", tends to imply the geographical characters in sea areas of Ningbo. The architecture form of "invisible" boat shape is derived from the "visible" ancient boat. It will be a culturalized morphology. Therefore, "visible" and "invisible" architectural forms naturally present "visible" and "invisible" cultural values. In spatial construction, the sunken entrance space which is derived from the model of the edge of "inland sea", exhibits the tension between space and time, and the connection with history and the past. Light-Shadow, virtual-real and dynamil-static comparisons in the spatial environment present elegant and harmonious spatial properties.

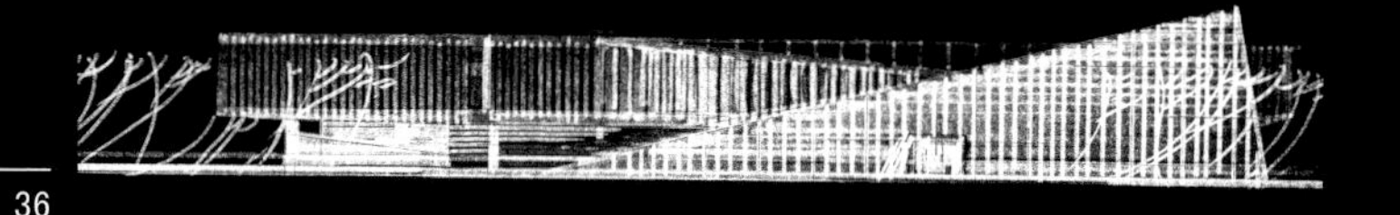

草图 | Sketch

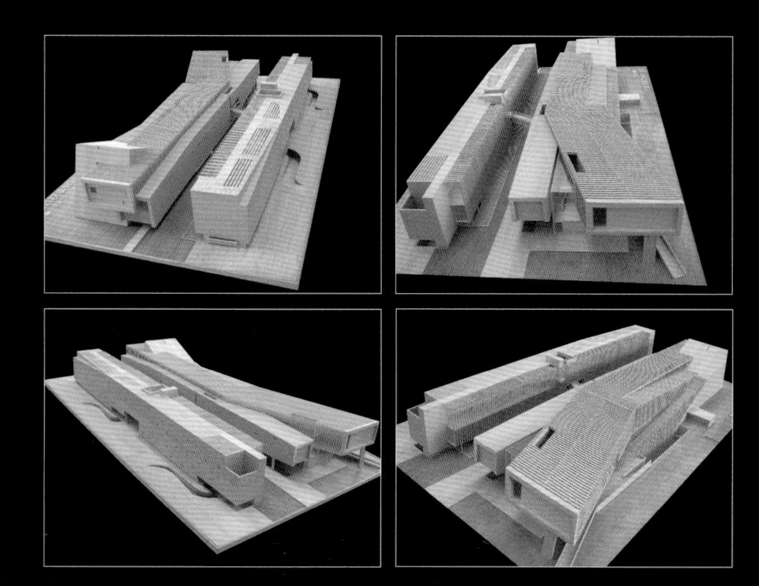

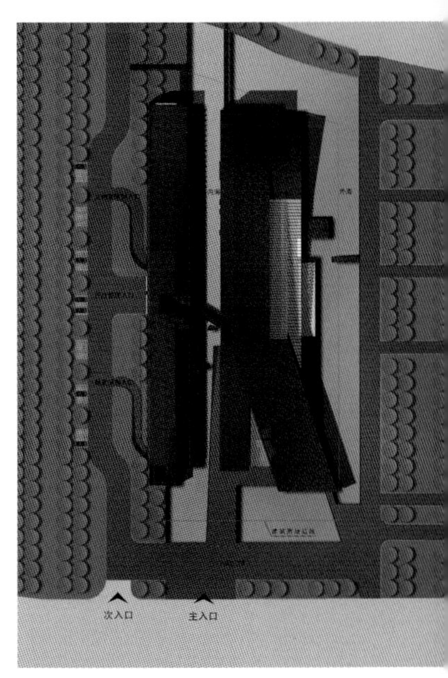

总平面图 | Site Plan

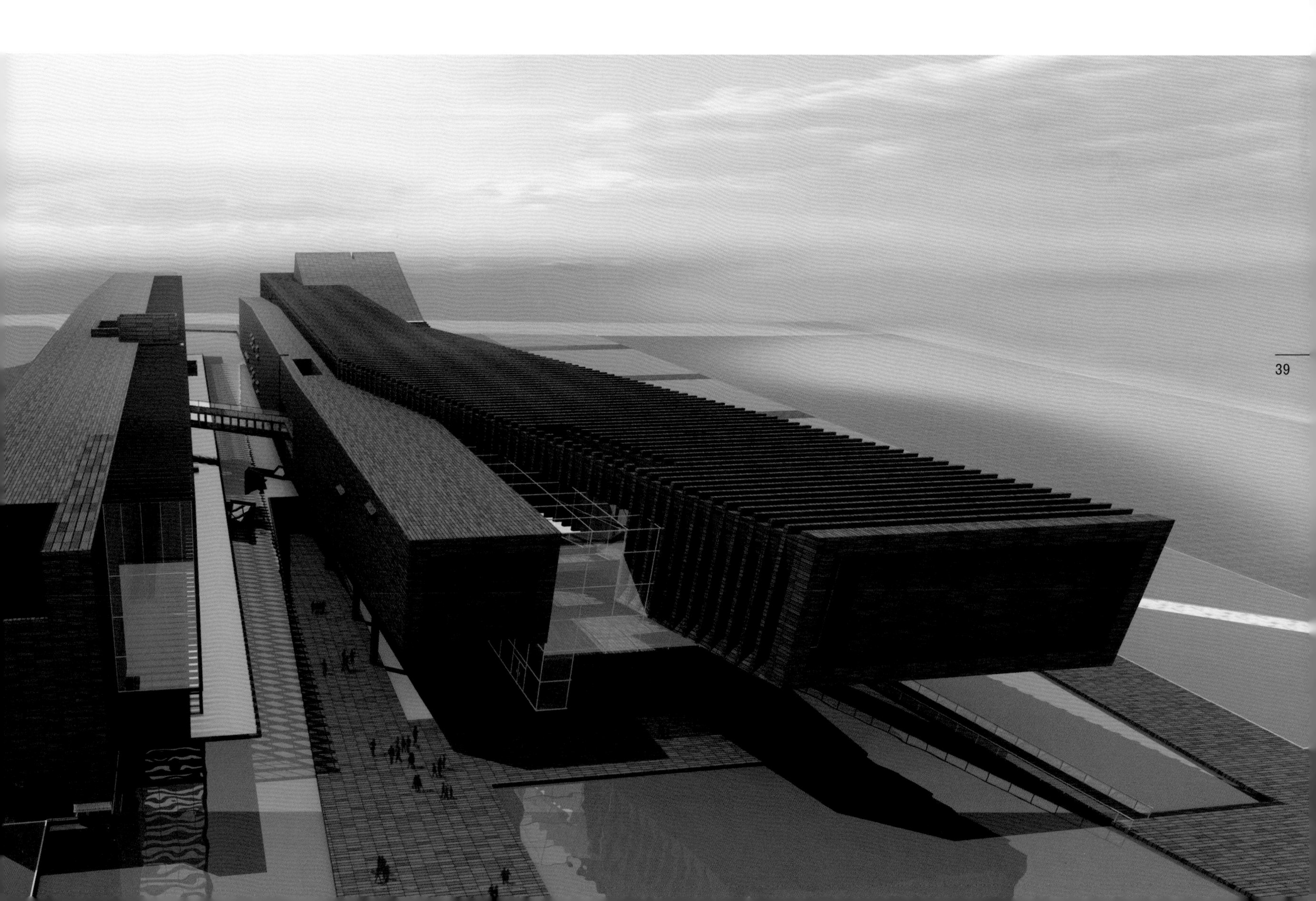

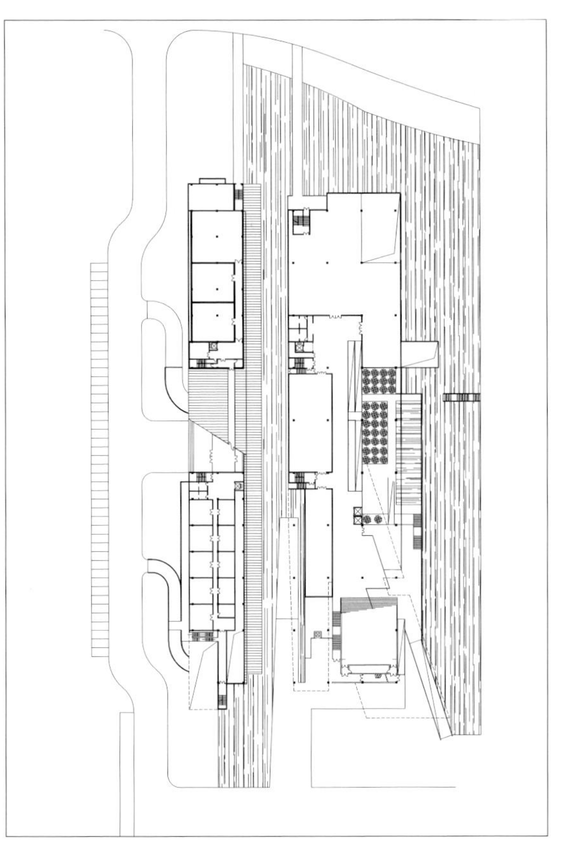

1.8 米标高层平面 | Plan at Level 1.8

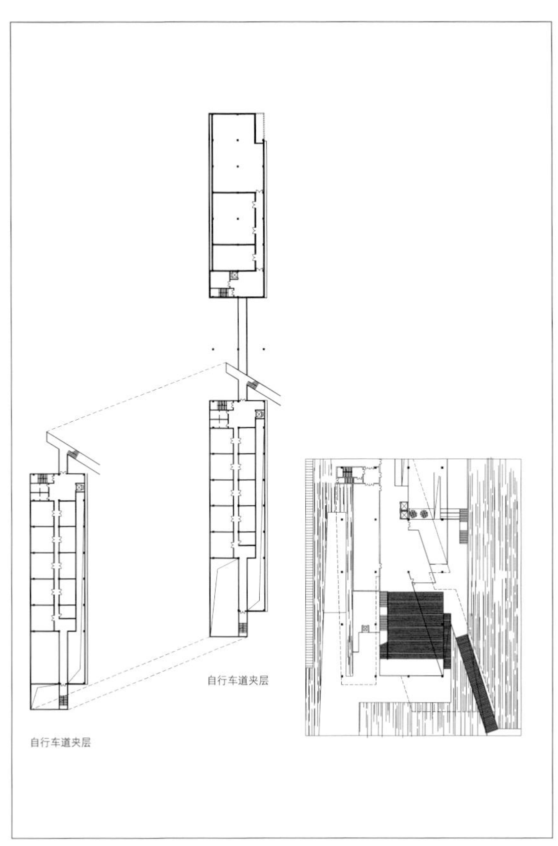

3.6/5.4/7.2 米标高层平面 | Plan at Level 3.6/5.4/7.2

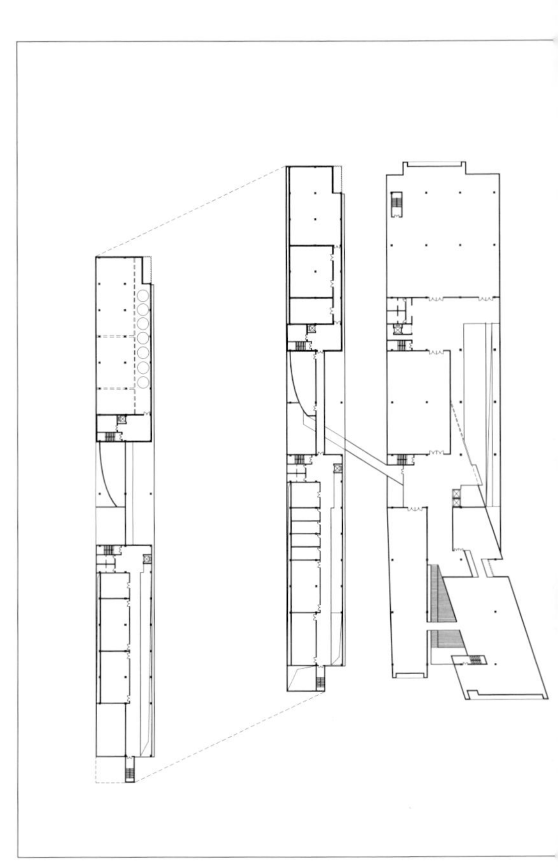

9/10.8/16.2 米标高层平面 | Plan at Level 9/10.8/16.2

内立面 1 | Internal Elevation1

内立面 2 | Internal Elevation2

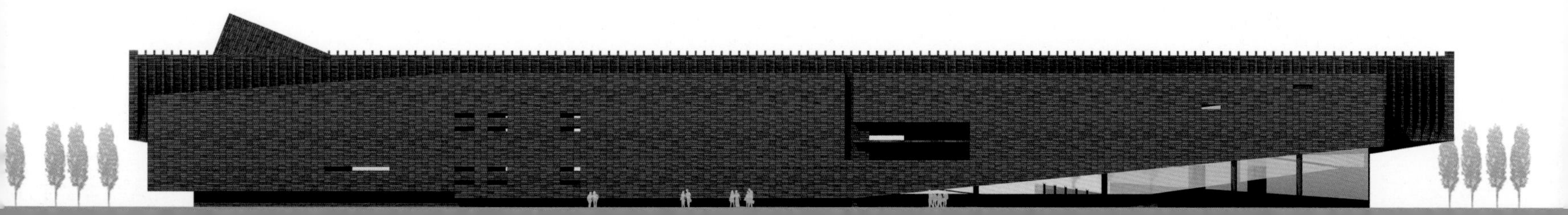

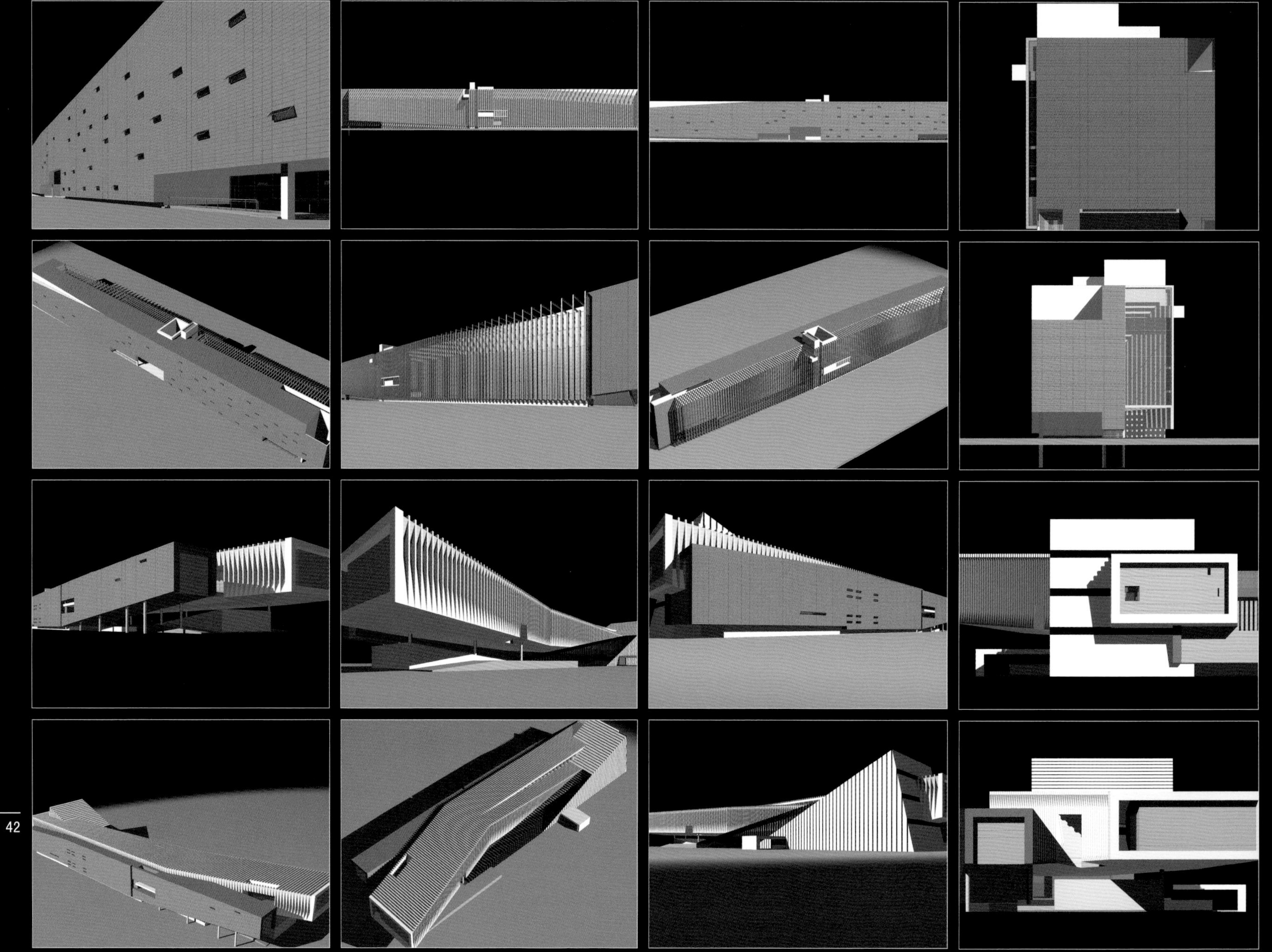

生成过程

东立面 | East Elevation

宁波鄞州金贸中心

Ningbo Yinzhou Trade Center

实 施 工 程　2005—2010
建筑师团队：高裕江、童辉、吴希良、章文杰
合 作 单 位：宁波中鼎建筑设计研究院

运用统一的设计手法，及大尺度的相似形体块，使整个建筑群北南二楼形成一高一低的态势，而又具和谐整体性。两楼共拥裙楼以期高度有机的融合正是设计的着力点之一。高效有机的平面构成，平直、方正的理性化形体架构，以及现代简约、典雅清新的形象塑造，体现出回归建筑本体的设计思想。

According to the application of unified design technique and arrangement of similar and large-scale spaces, the North and South buildings form a relationship of "one high and another low" and the whole building complex comes to be harmonious and integrated. The organic combination of the two buildings which share the common podium is the key point for the design. Efficient and organic plane composition, straight and square spatial construction, and modern and elegant image creation all tend to present the design idea of renaissance of ontological architecture.

总平面图 | Site Plan

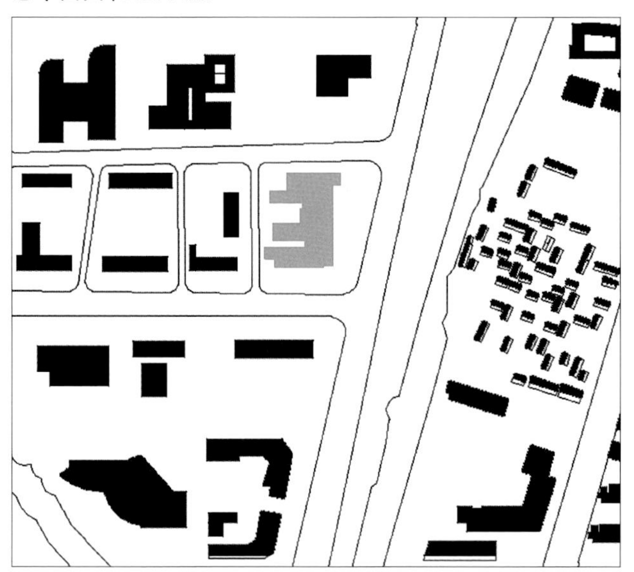

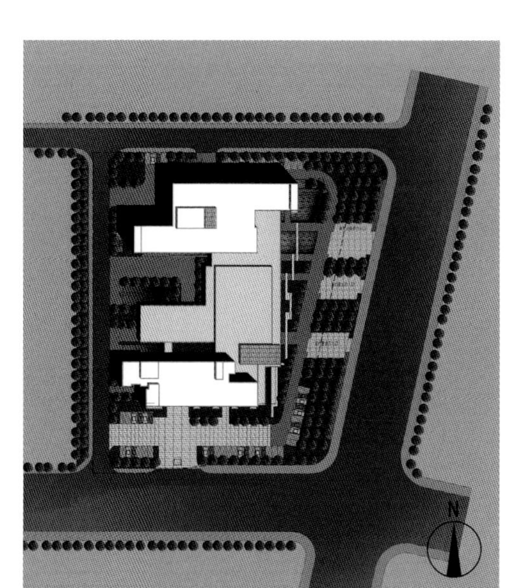

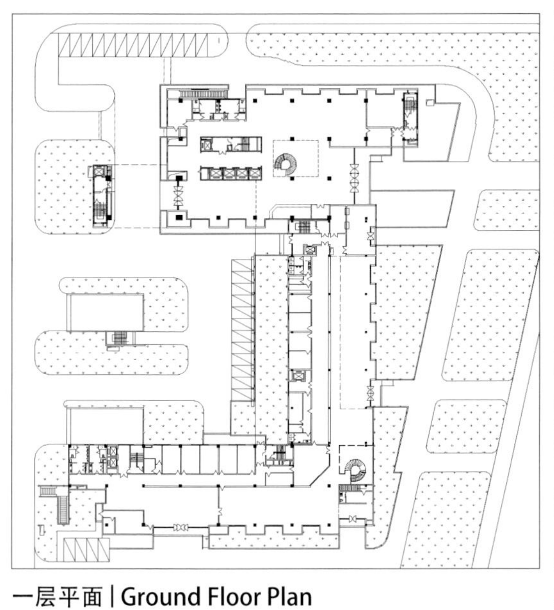

一层平面 | Ground Floor Plan

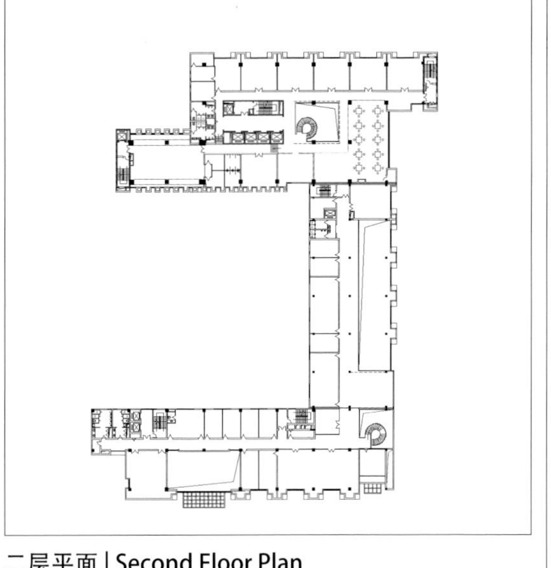

二层平面 | Second Floor Plan

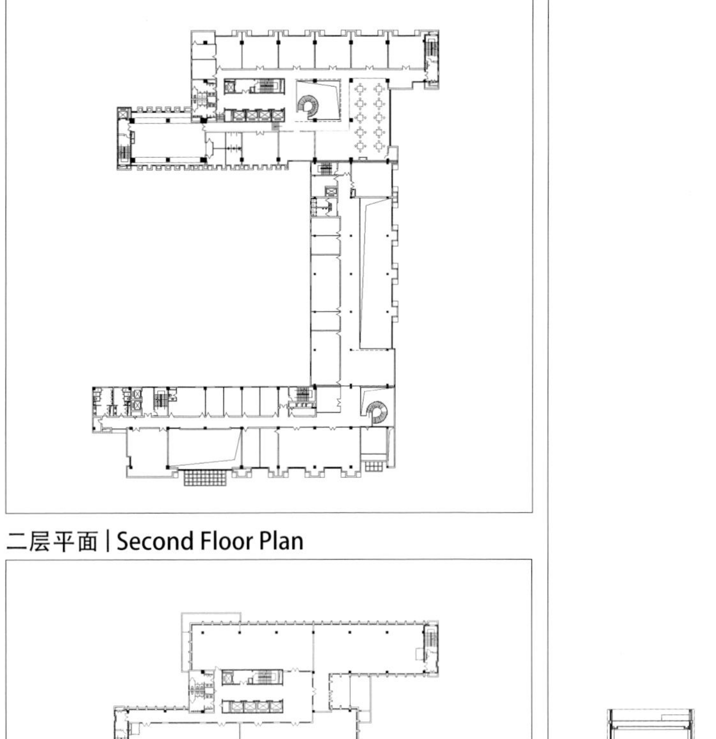

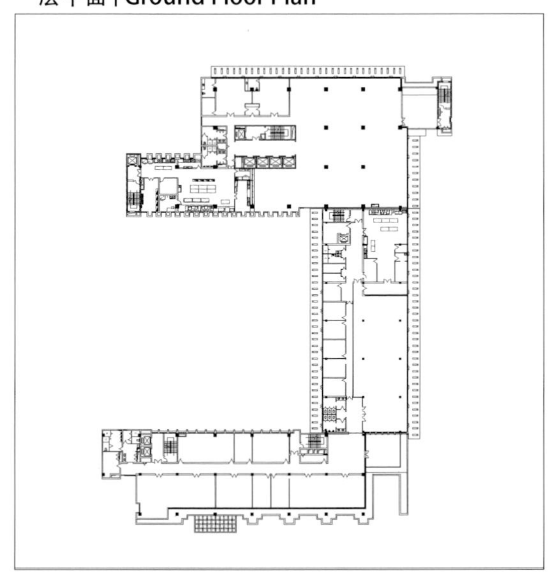

三层平面 | Third Floor Plan

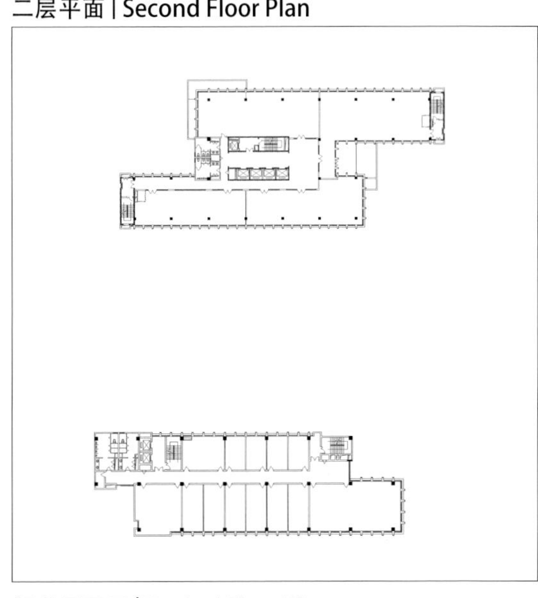

标准层平面 | Typical Floor Plan

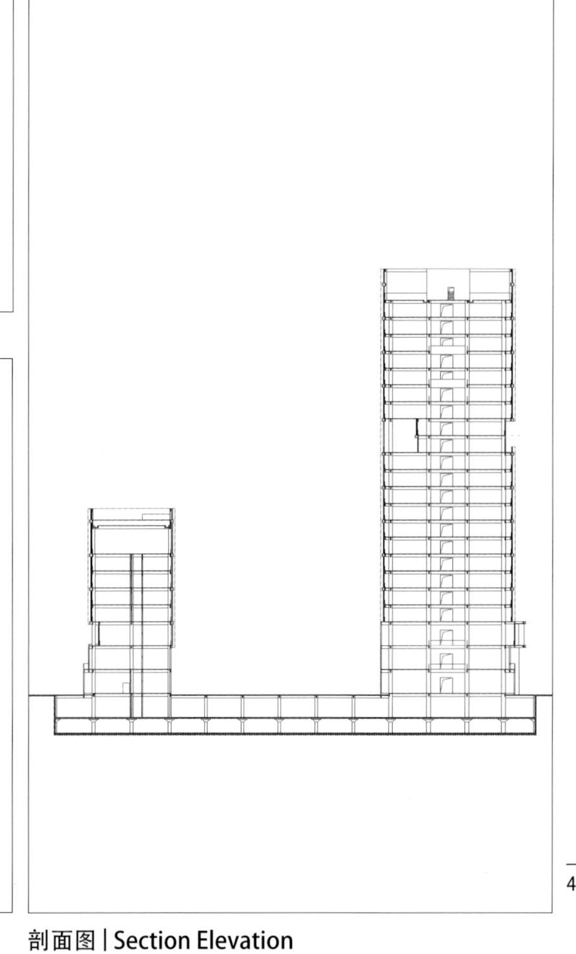

剖面图 | Section Elevation

新洲银座

宁波奥克斯总部大楼

Ningbo AUX Headquarter

实 施 方 案　国际竞标（第一名）/ 2006—2012
建筑师团队：高裕江、胡慧峰、章文杰、卢挺、高生皓、
程锦、段威、李航

形似“X”的形体组构呼应城市设计的目标：本楼夹于东西超高层大楼之间，设计采用“折板互动”“两板成塔”的建筑形式，能较好地解决内在生态绿色景观需求与外部城市天际线、空间景观的呼应关系。适度植入和营建绿色生态的中庭空间，提升商务办公环境品质，是一种可取的设计策略。

The X-shaped spatial configuration is directly related to urban design: the construction is in the middle of the east and west high-rise buildings, adopting architectural forms of "interaction between folded-plates" and "tower formed by two panels" to solve problems concerning the corresponding relationship between the needs of the internal green ecological landscape, and the external city skyline and landscape. In order to improve the quality of the environment of business offices, it should be a practical strategy to build up green ecological atrium spaces.

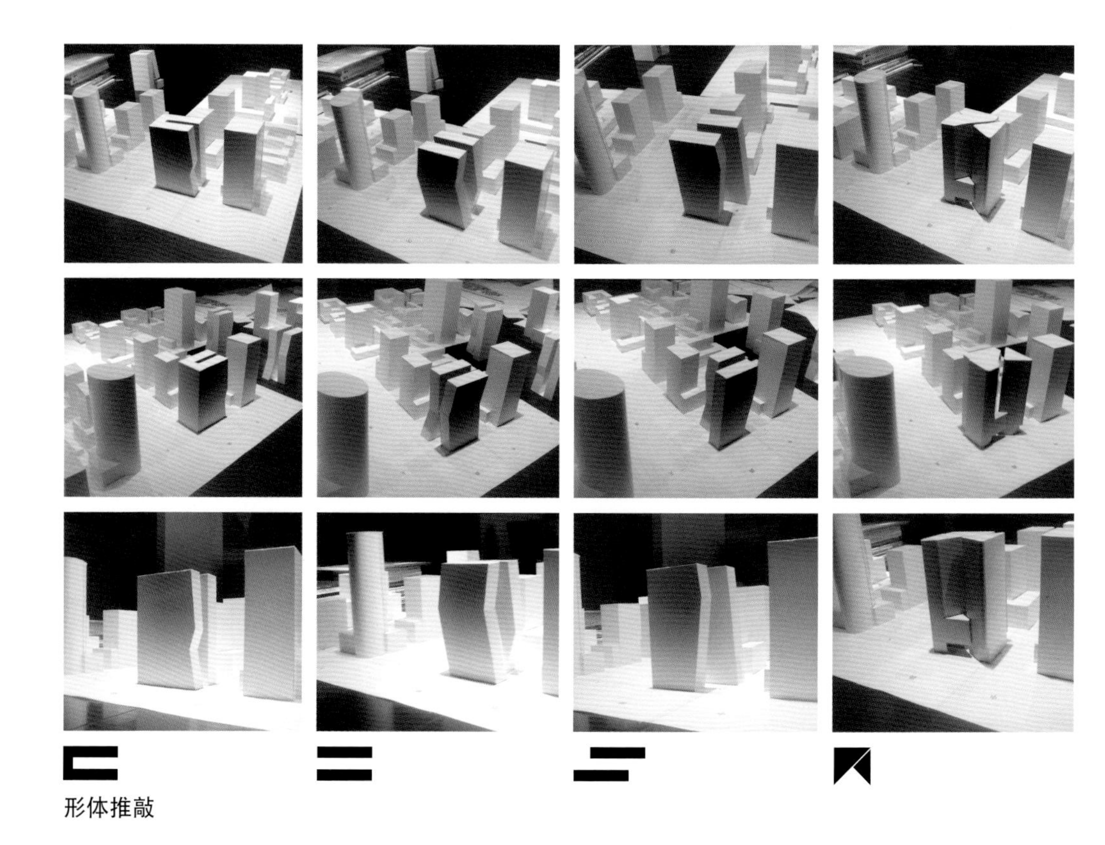

形体推敲

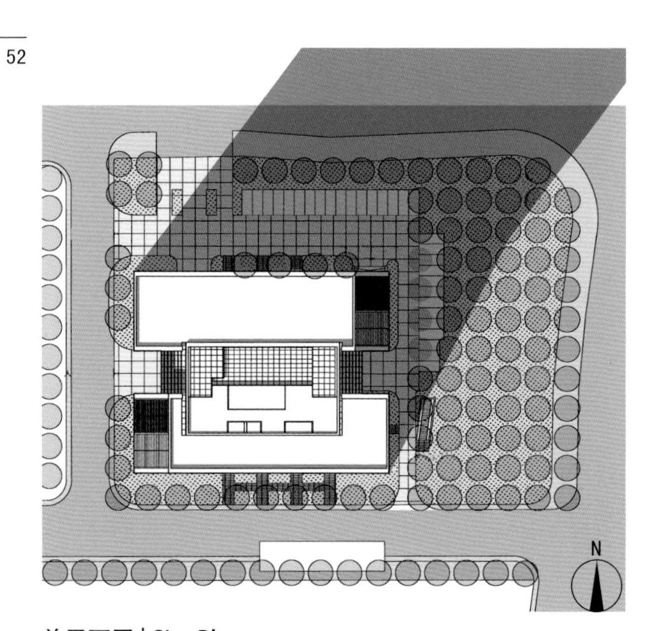

总平面图 | Site Plan

AUX

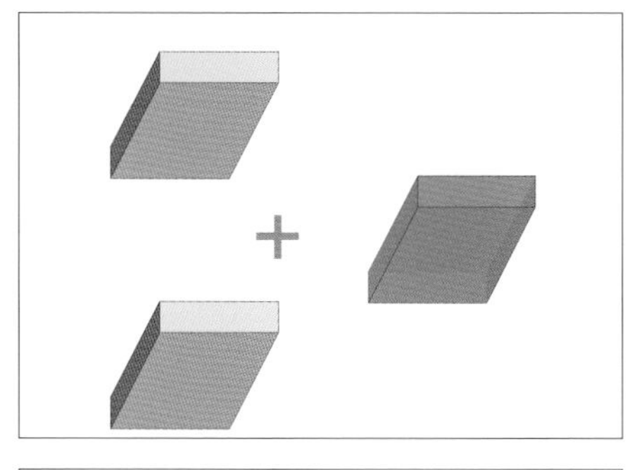

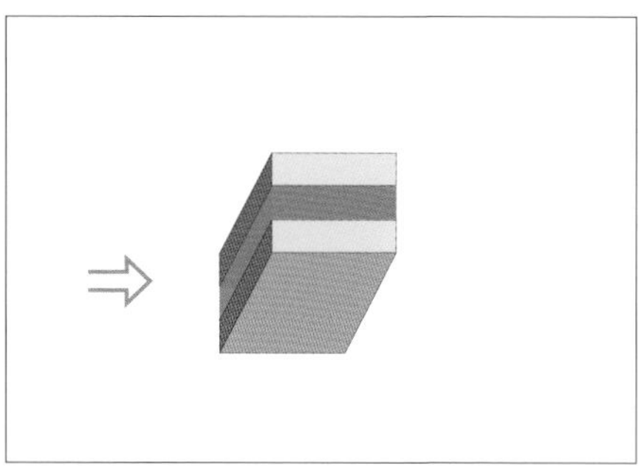

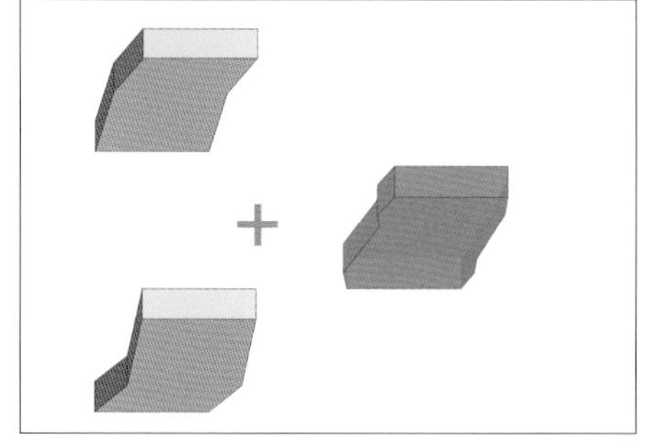

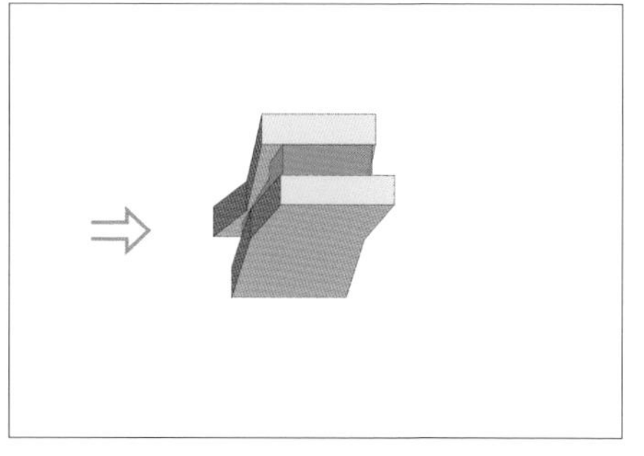

形态生成

中庭空间分析

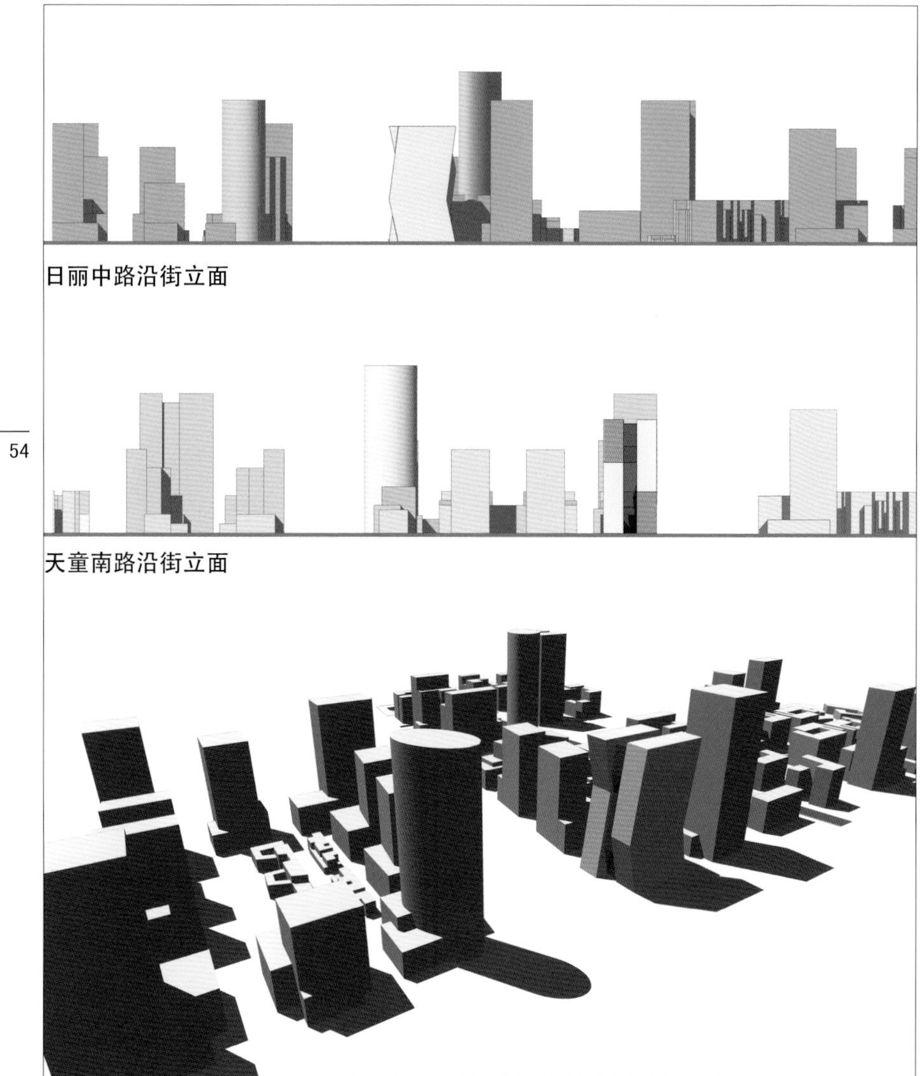

日丽中路沿街立面

天童南路沿街立面

建造环境分析

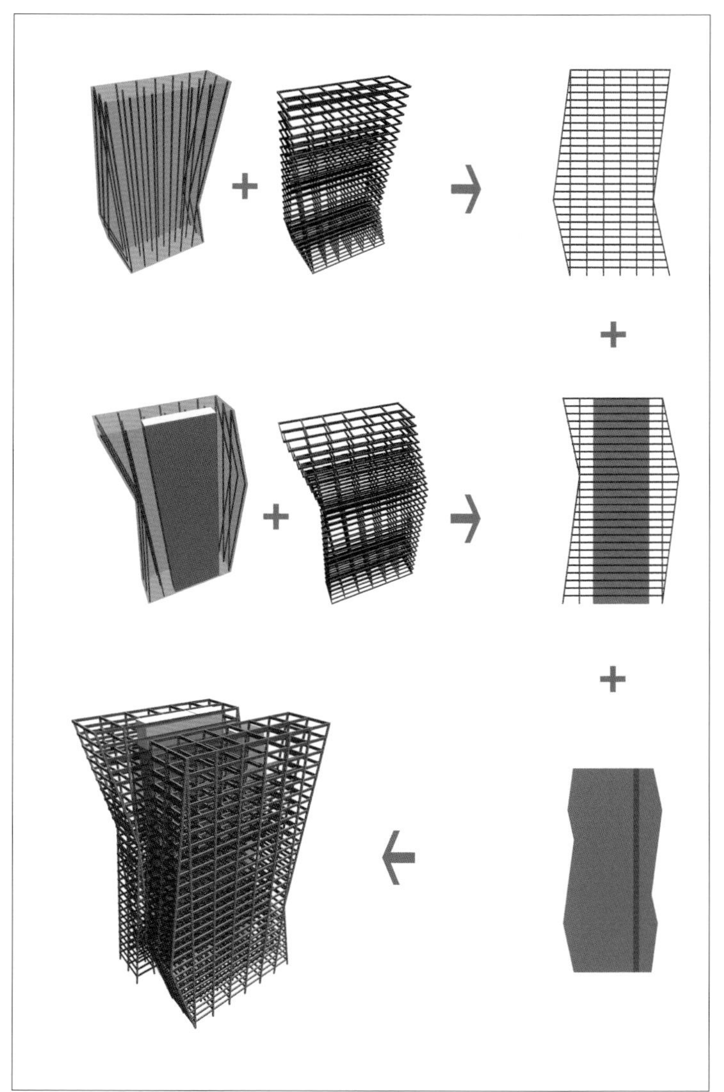

结构分析

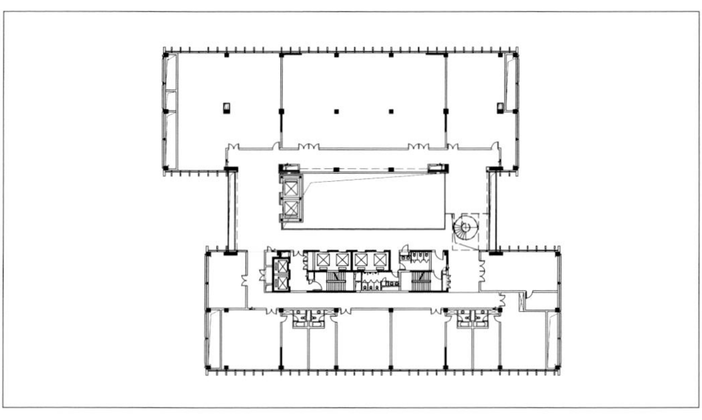
二十六层平面 | Twenty-sixth Floor Plan

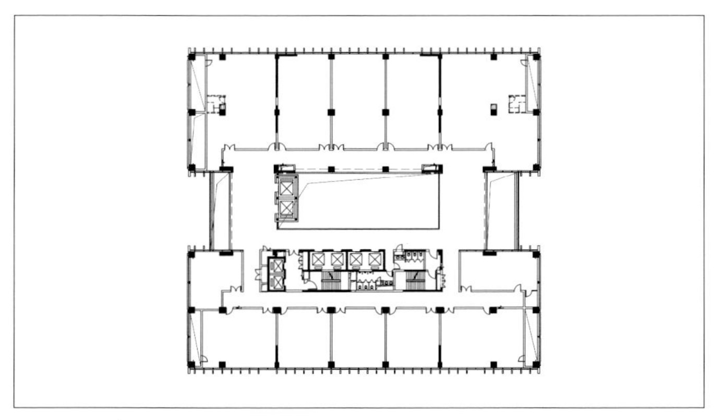
七层平面 | Seventh Floor Plan

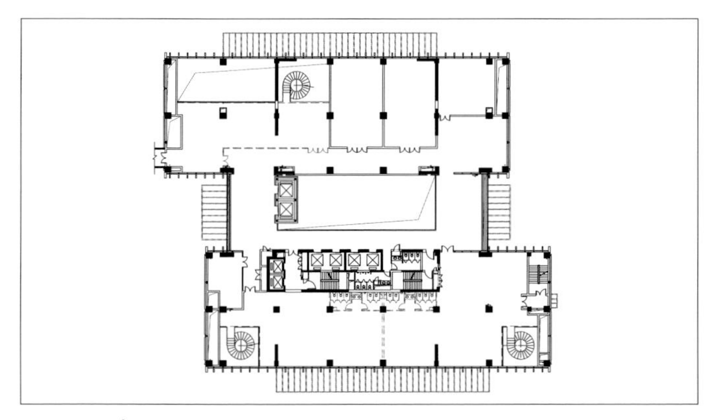
二层平面 | Second Floor Plan

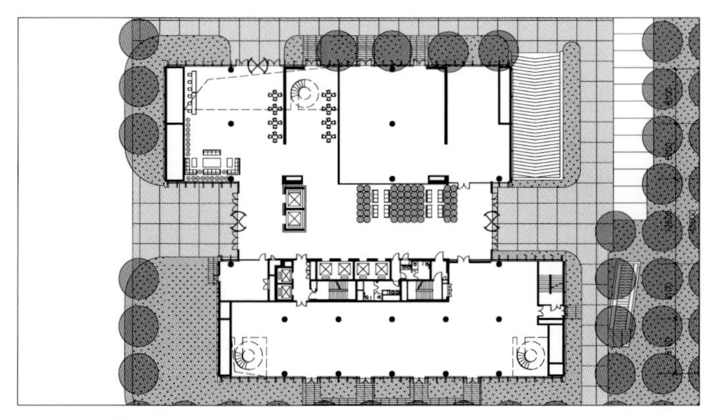
一层平面 | Ground Floor Plan

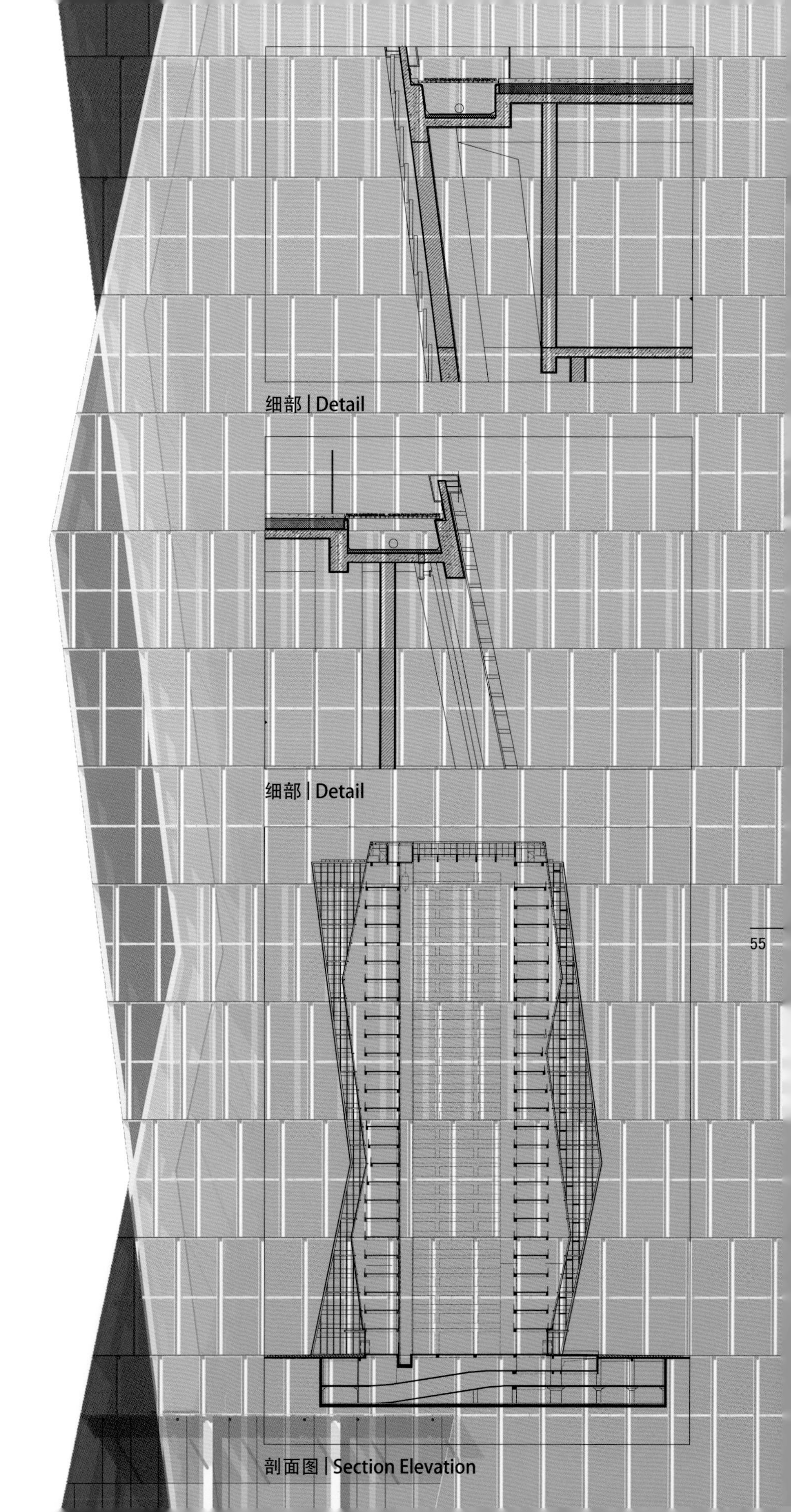
细部 | Detail

细部 | Detail

剖面图 | Section Elevation

AUX

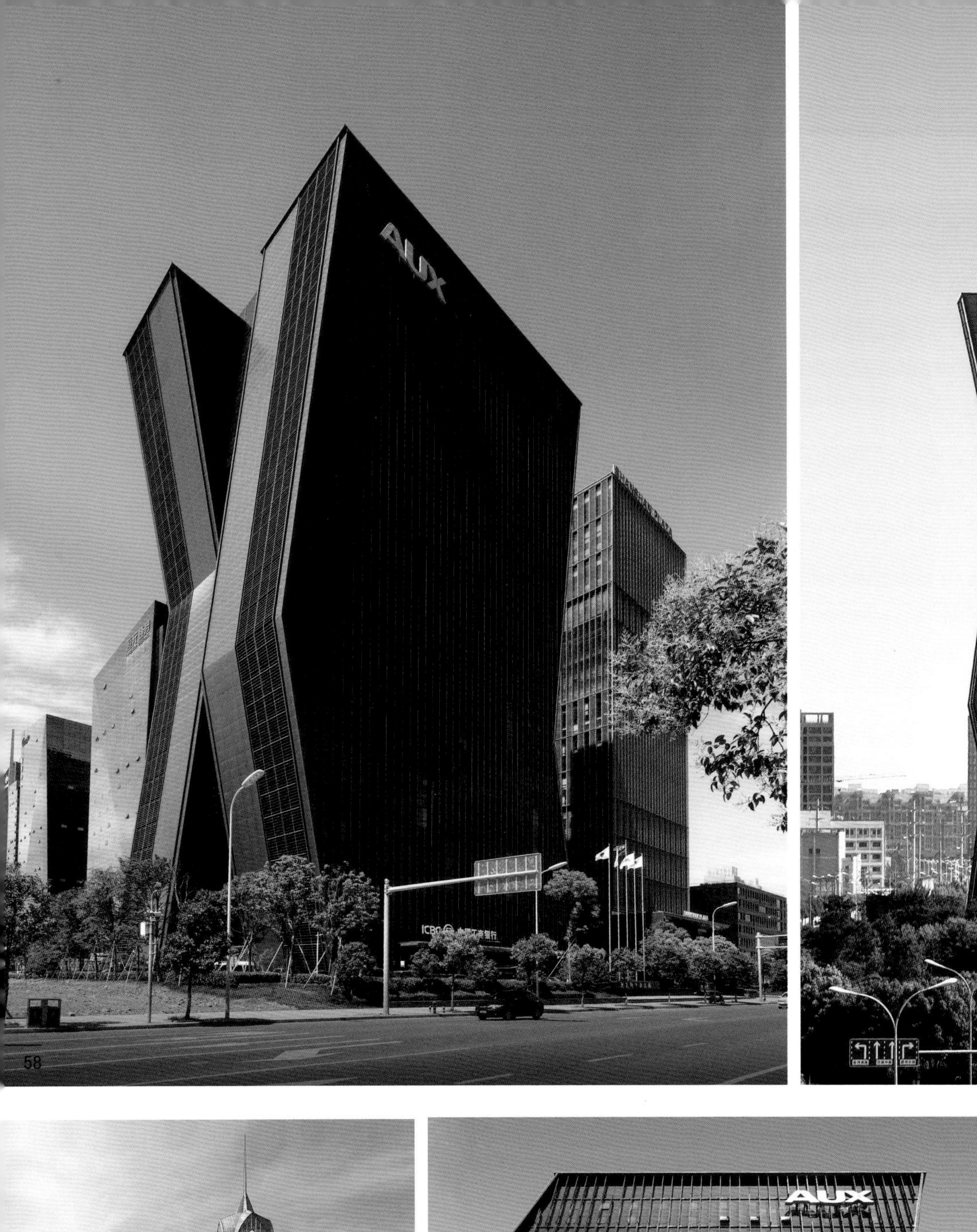
AUX

AUX

AUX

AUX

AUX

AUX

江苏美术馆

Jiangsu Art Gallery

设 计 方 案　国际竞标 /2006
建筑师团队：高裕江、卢挺、章文杰、高庶三、高生皓

源自中国传统印鉴——方印的造型，以及出自国文综艺体“艺”的数字化处理后的窗格体表面肌理的有机整构，通过倾斜、扭动的动势整合，一座现代而古悠、简约而意趣的城市标志性建筑呼之而出。它是建筑化的雕塑，也是雕塑化的建筑。俗称“南京红”的暗红色调，具有强烈的地域文化色彩以及浓郁的人文精神气息。

Derived from the traditional Chinese seal——the model of square seal, and the organic reconstruction of window screen's volume-surface texture as a consequence of the digitalization of Chinese Mixed Art Font "Yi", through the integration of motive forces: sloping and twisting, a city's symbolic architecture which is modern and ancient, simple and intersting is vividly portrayed. It is not only an architectural sculpture but also a sculpturalized architecture. The dark red commonly called as "Nanjing Red", features intense regional cultural characters and distinctive spirit of humanism.

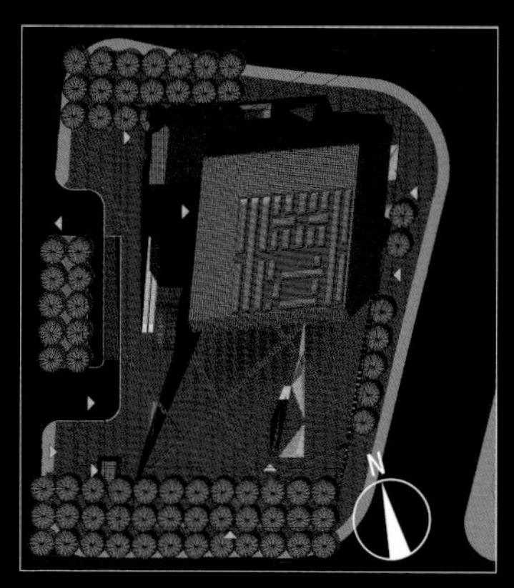

总平面图 | Site Plan

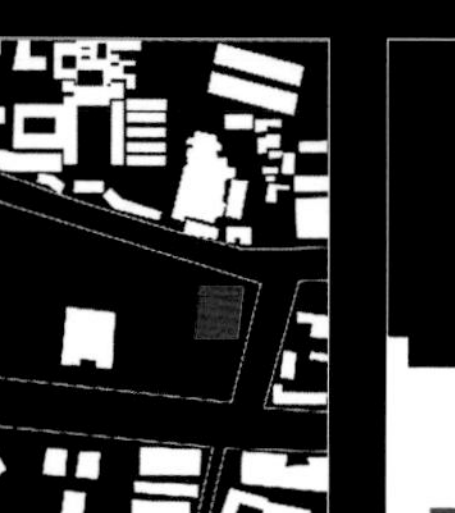

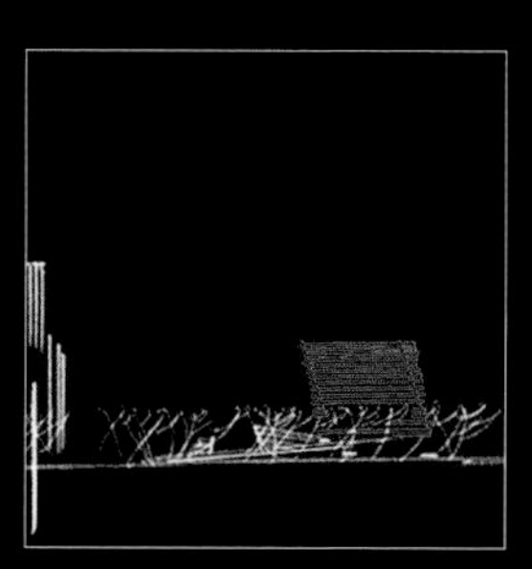

概念构思

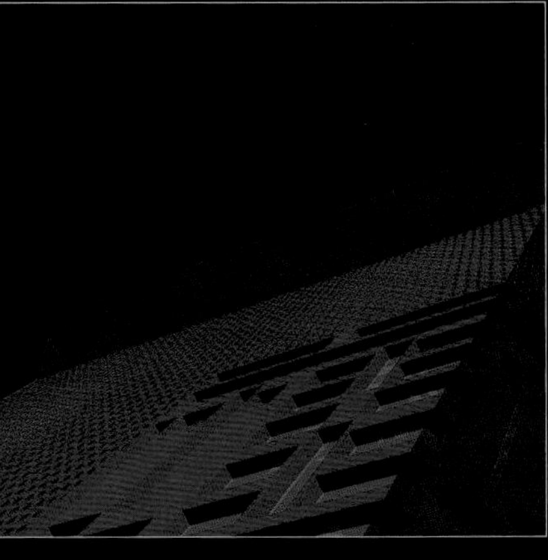

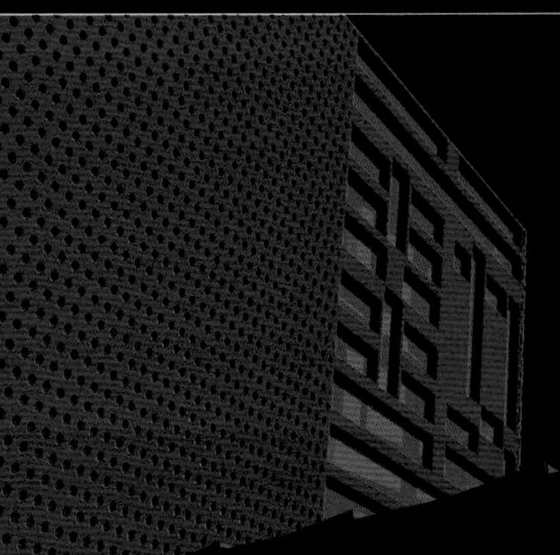

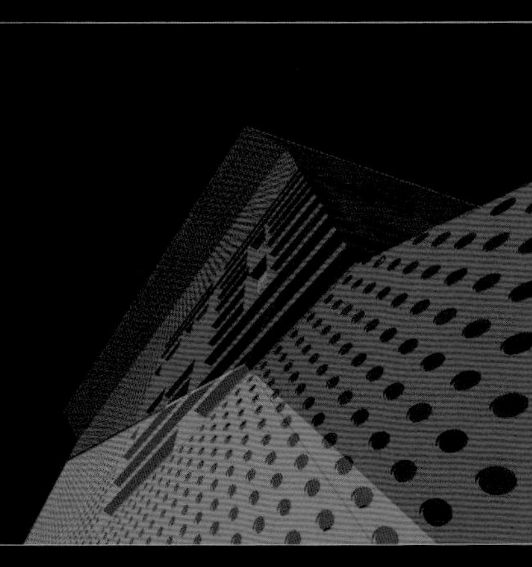

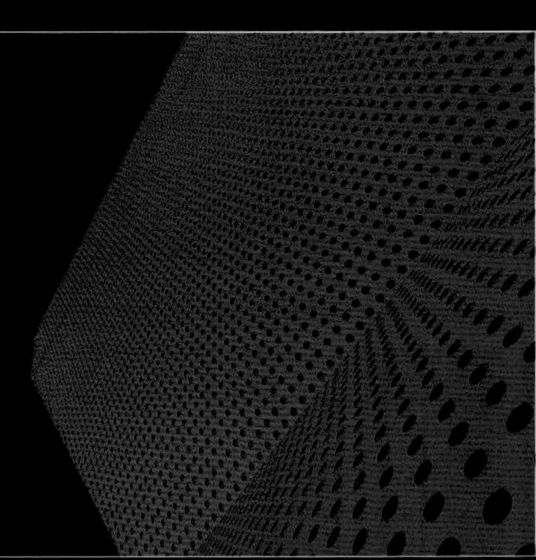

六层平面 | Sixth Floor Plan

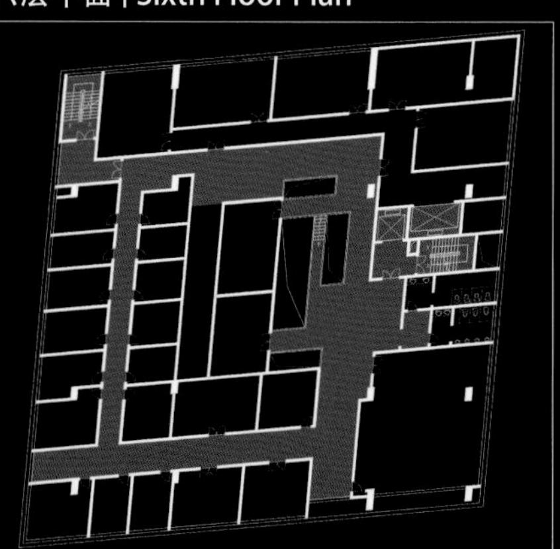

五层平面 | Fifth Floor Plan

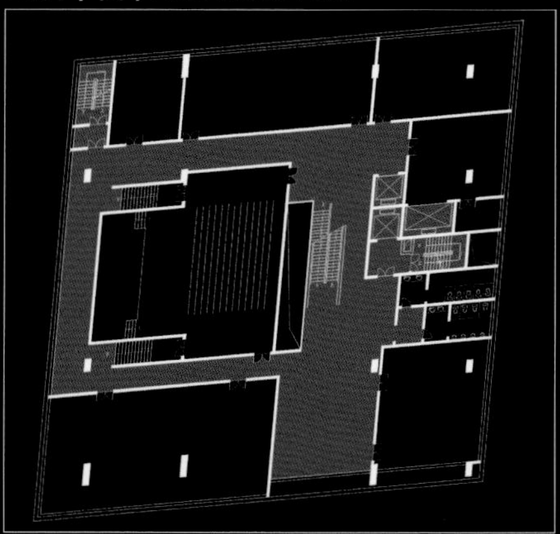

四层平面 | fourth Floor Plan

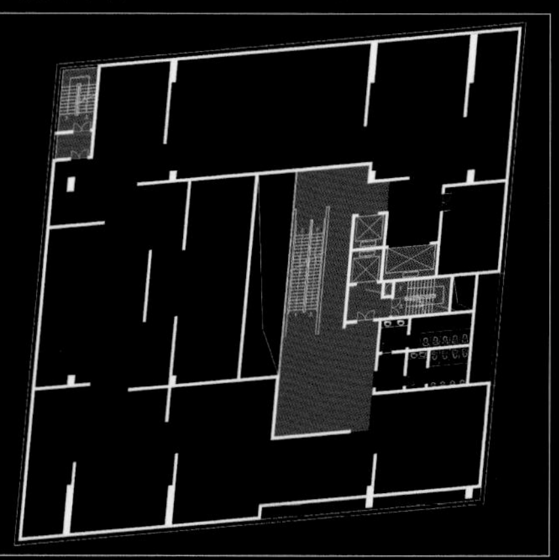

三层平面 | Third Floor Plan

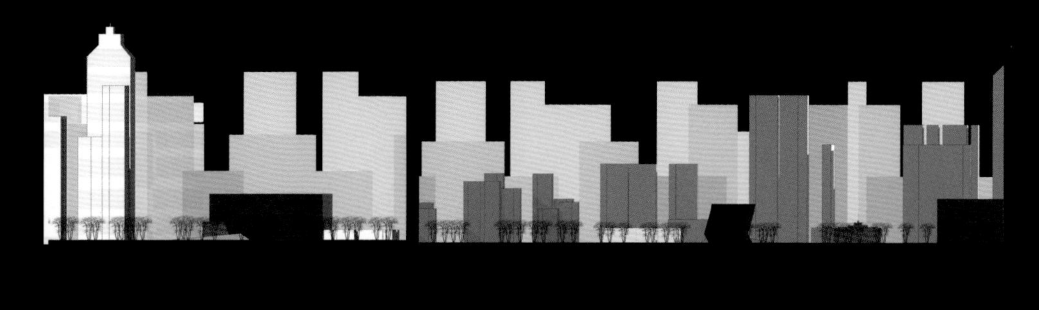

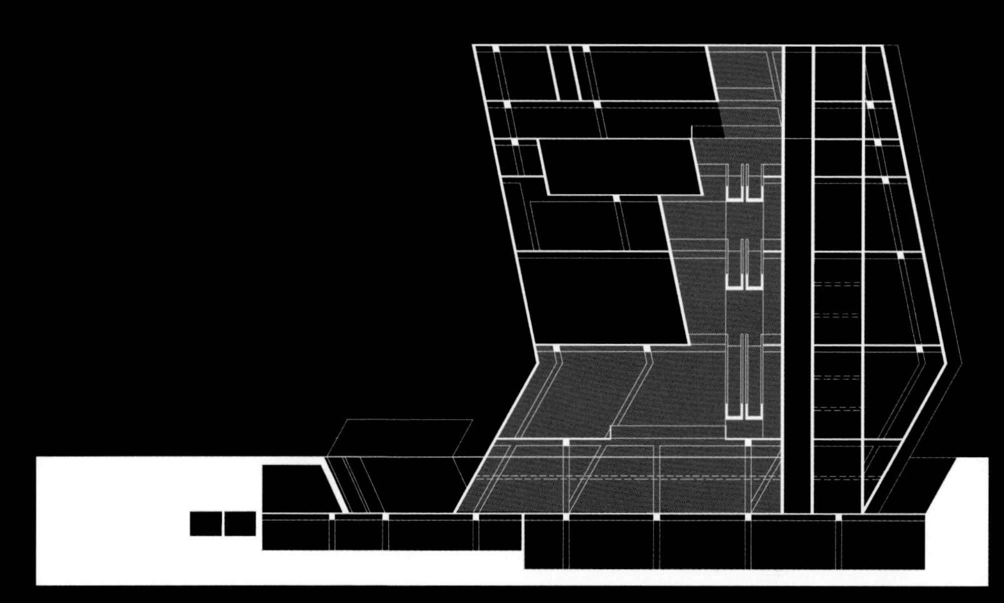

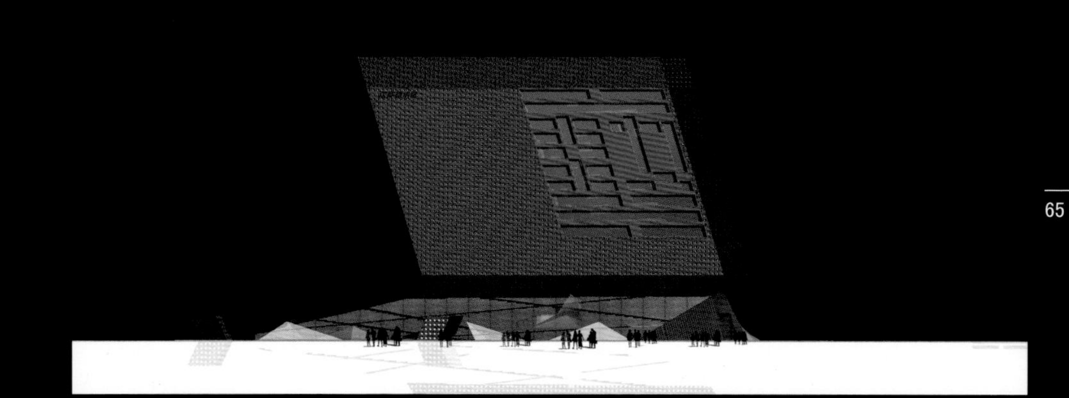

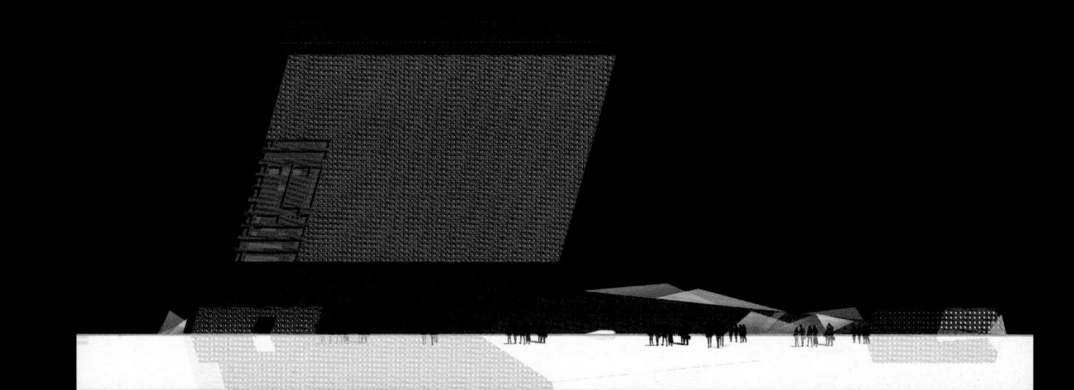

宁波布利杰总部大楼

Ningbo Bulijie Headquarter Design

设 计 方 案　国际竞标 /2006
建筑师团队：高裕江、章文杰、高庶三、卢挺、高生皓

处于高楼林立中的布利杰总部大楼设计通过雕塑化的形体策略，以及细孔金属百叶网片界面肌理的生态化构筑，在城市空间环境层面和企业内在文化展示中，进行有机整合，突显其特色和个性。内部中庭共享空间注重生态宜人、景观互动的环境效果，同时体现办公空间的理性、高效、便捷的建筑类型学特征。

Bulijie Headquarter located in the block of a myriad of high-raise buildings. Through the sculpturalized form and the ecological construction of the perforated metal grill interface, the urban space environment and its enterprise culture are organically combined together to highlight the company's characteristics. The inside shared-courtyard emphasizes ecological and dynamic environment, and meanwhile , it tends to present a rational, efficient and convenient architectural style.

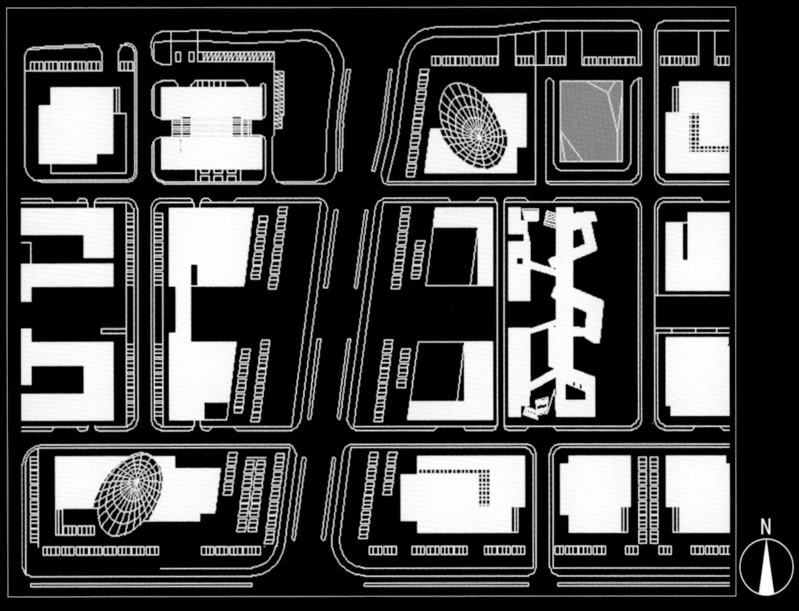

总平面图 | Site Plan

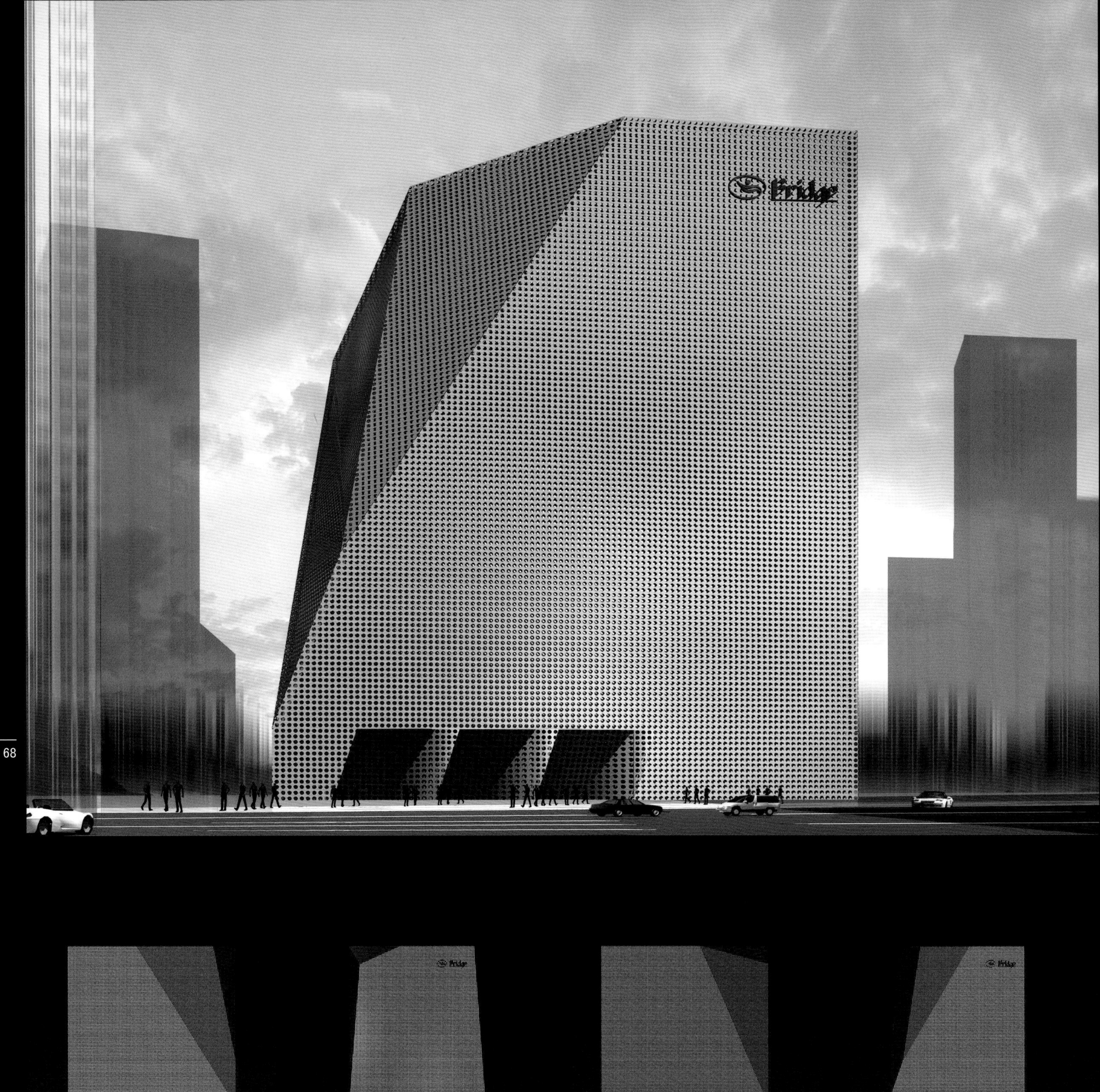

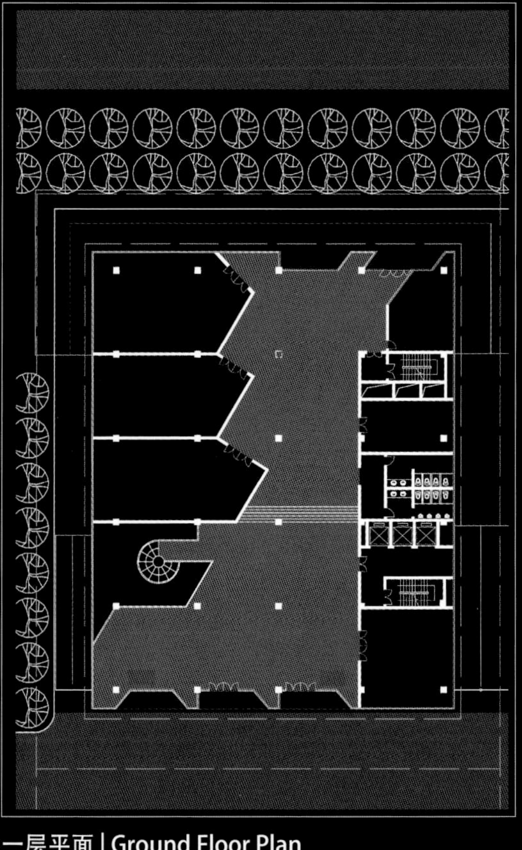

一层平面 | Ground Floor Plan

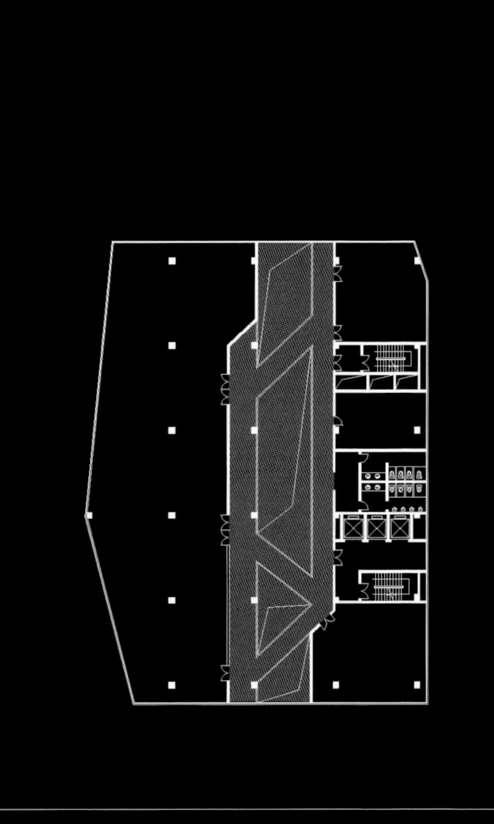

标准层平面 | Typical Floor Plan

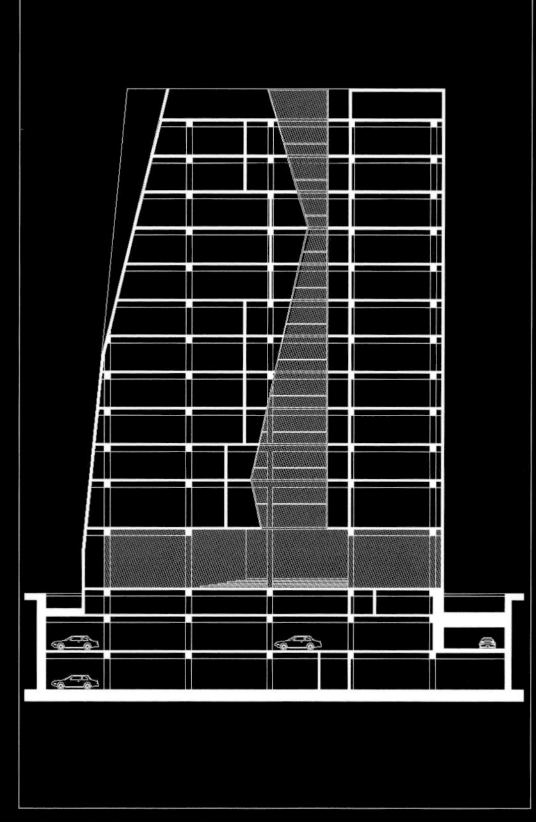

剖面图 | Section Elevation

方案比较

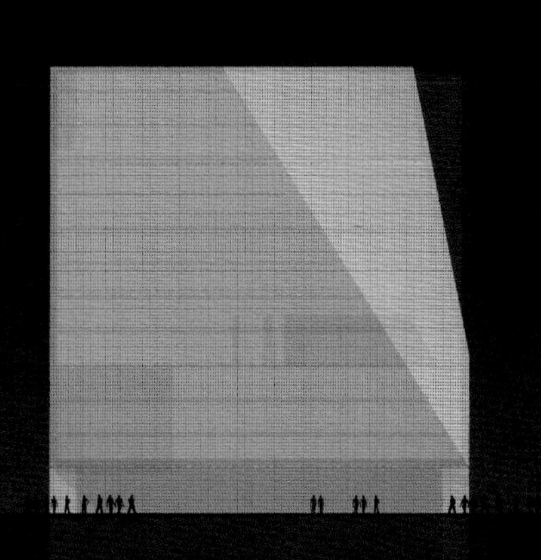

南浔行政综合办公楼

Nanxun Administrative Office Buildings

概 念 方 案　国内竞标 /2006
建筑师团队：高裕江、章文杰、曹晓敏、夏琴、邓飞、包峥

设计通过多重传统院落型的空间模式，使建筑群整体合一，强化生态节能的同时承延传统民居建筑的空间文化特色，结合绿化水系及滨水风貌，突显江南水乡意韵。建筑界面大胆采用黑白水墨的色彩系列，建筑形体形象源自南浔民居“观音兜”的形式，使建筑呈现出鲜明时代性和地域文化性特色。

Through multiple traditional courtyard space model, the design integrates the buildings, strengthening ecological and energy saving ,at the same time, inheriting building space and cultural characteristics of traditional local-style dwelling houses, combined with the green river and waterfront landscape, highlighting the water brush of regions south of the Yangtze River. The building interface adopt the color of black and white ink series and the building forms come from Nanxun local-style dwelling houses form of "guanyin bucket", which present the distinct era and regional cultural features.

民居分析

总平面图 | Site Plan

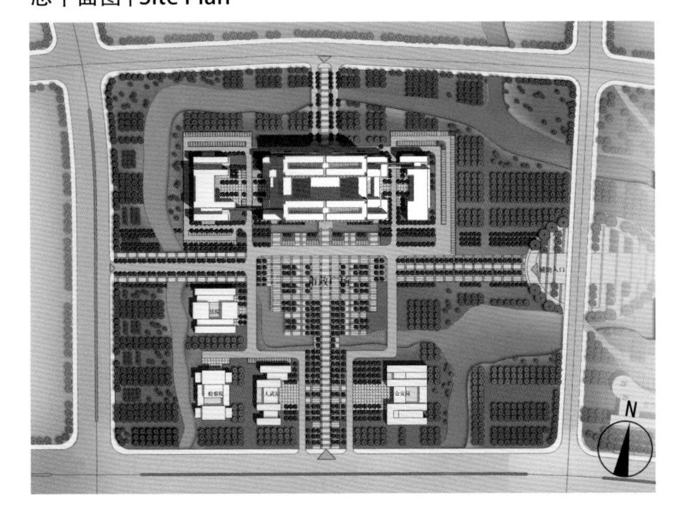

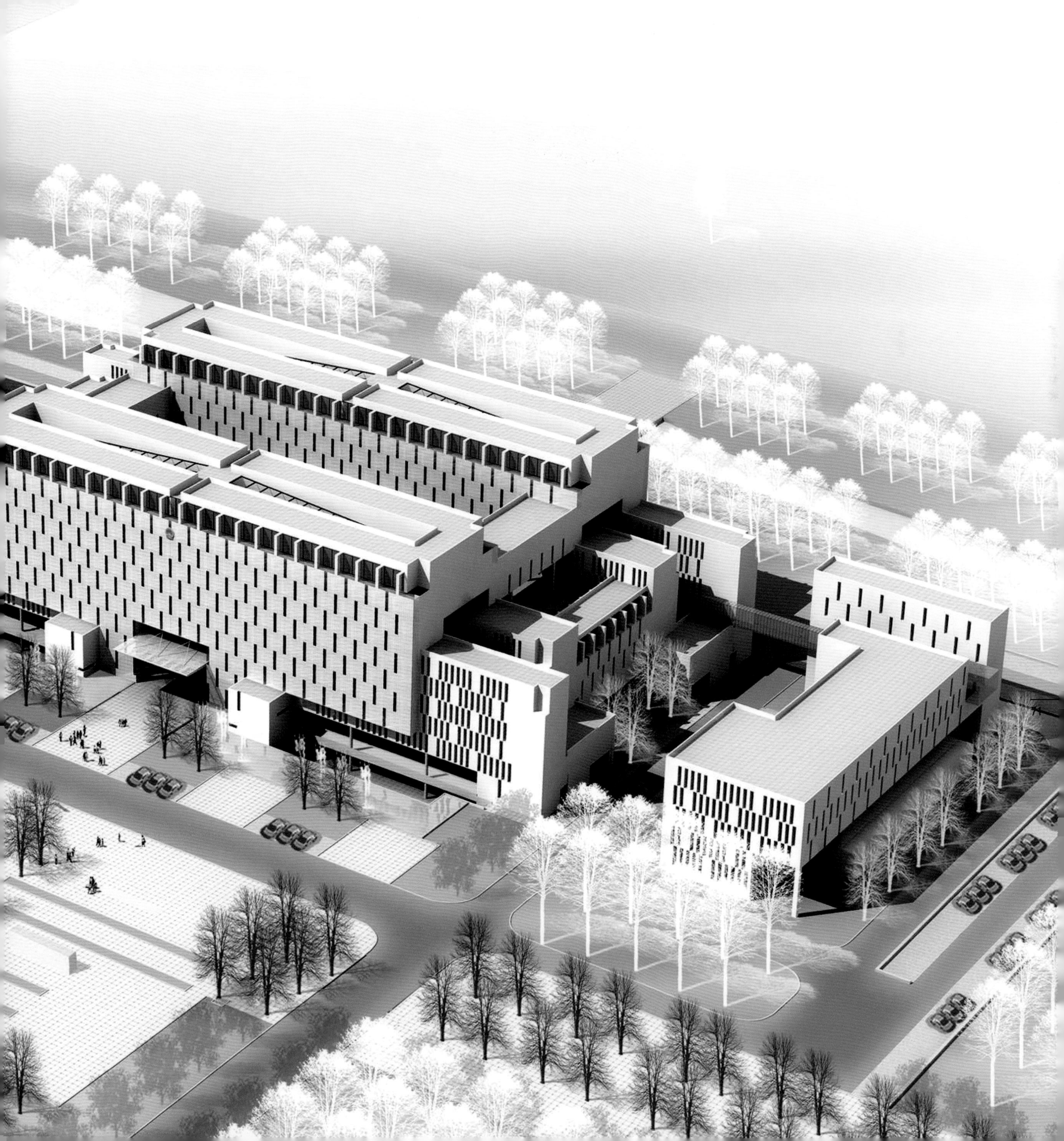

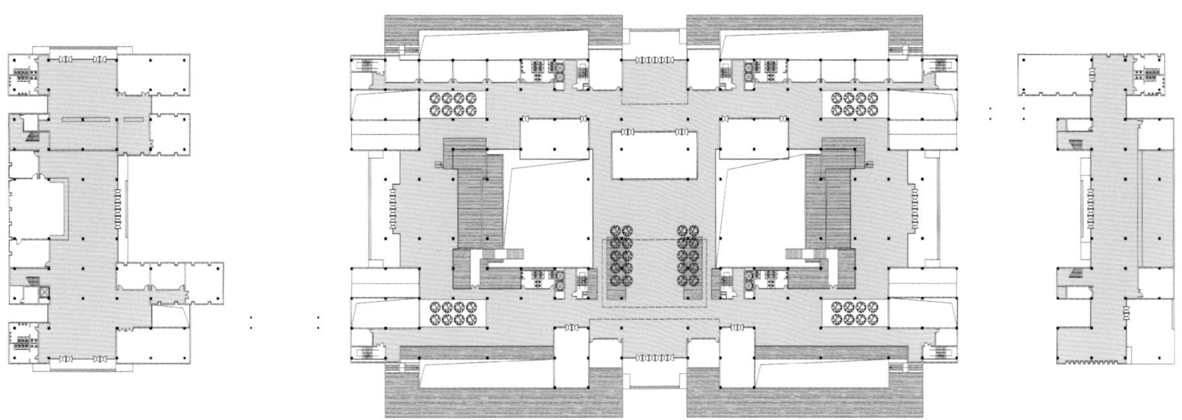

一层平面 | Ground Floor Plan

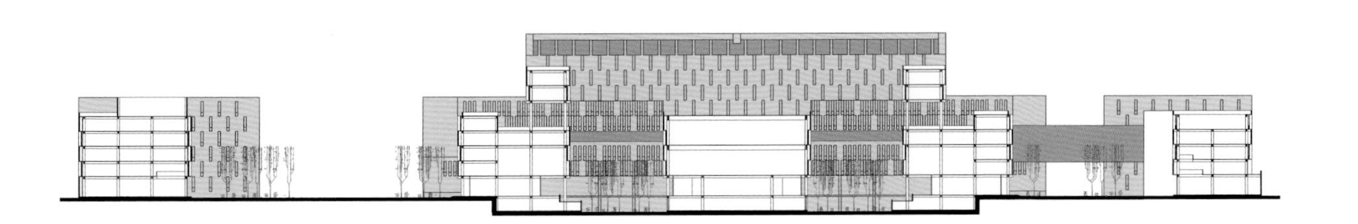

剖面图 | Section Elevation

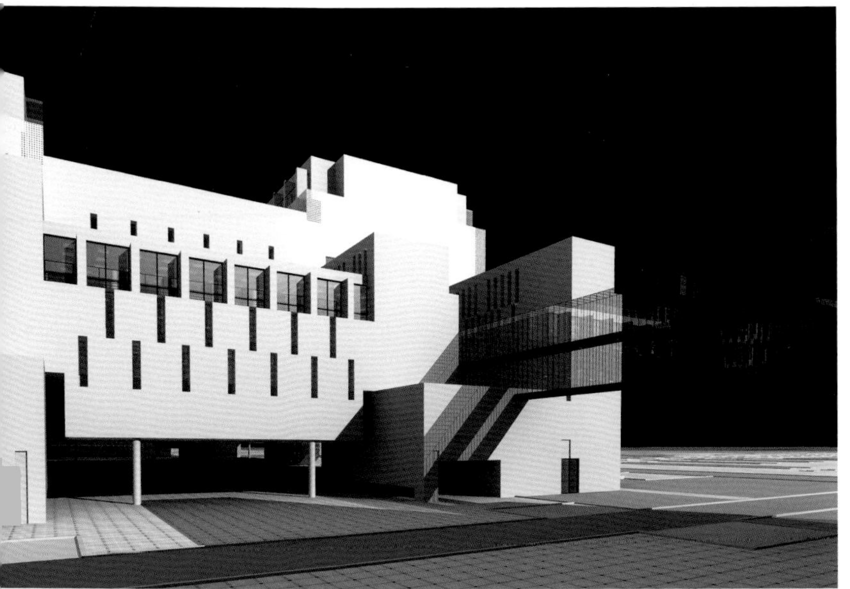

宁波鄞州金融中心

Ningbo Yinzhou Financial Center

概 念 方 案　2007
建筑师团队：高裕江、章文杰、秦浩、林萍英

在简约理性、现代大气、实用经济为基本设计原则下，设计通过强化城市、建筑、环境各层面要素的整构与互动，达到整体和谐的效果。单体建筑矩形错位的平面与竖向退台的收缩有机构筑，形成清新现代、挺拔高矗、富有动势和雕塑感的形体。同时，金属构件及节能幕墙的组合运用，增添了建筑的时代性、技术性和艺术性。

Due to fundamental design principle of achieving concise, rational, modern, practical and economic architecture, the reconstructions and interactions of city, building and surroundings are reinforced to create a harmonious environment as a whole. The organic combination of heterogeneous rectangular architectural plan and vertical terraces which are stepping backwards forms the fresh, modern, tall, straight, dynamic and sculptural architecture. Meanwhile, the metal components and energy saving curtain wall are assembled together to present the architectural characters of time, technological and artistic values.

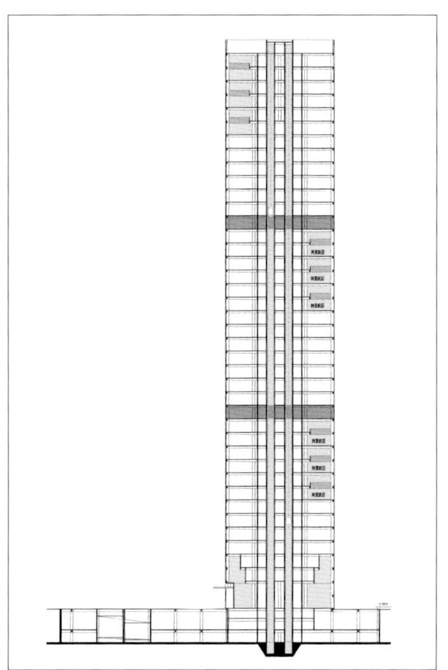

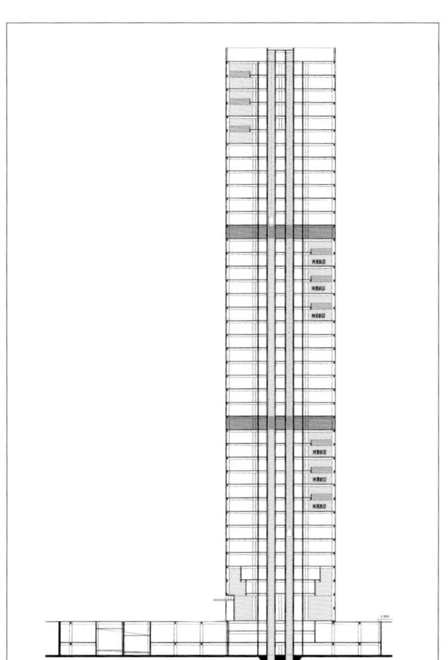

剖平面 | Section Elevation

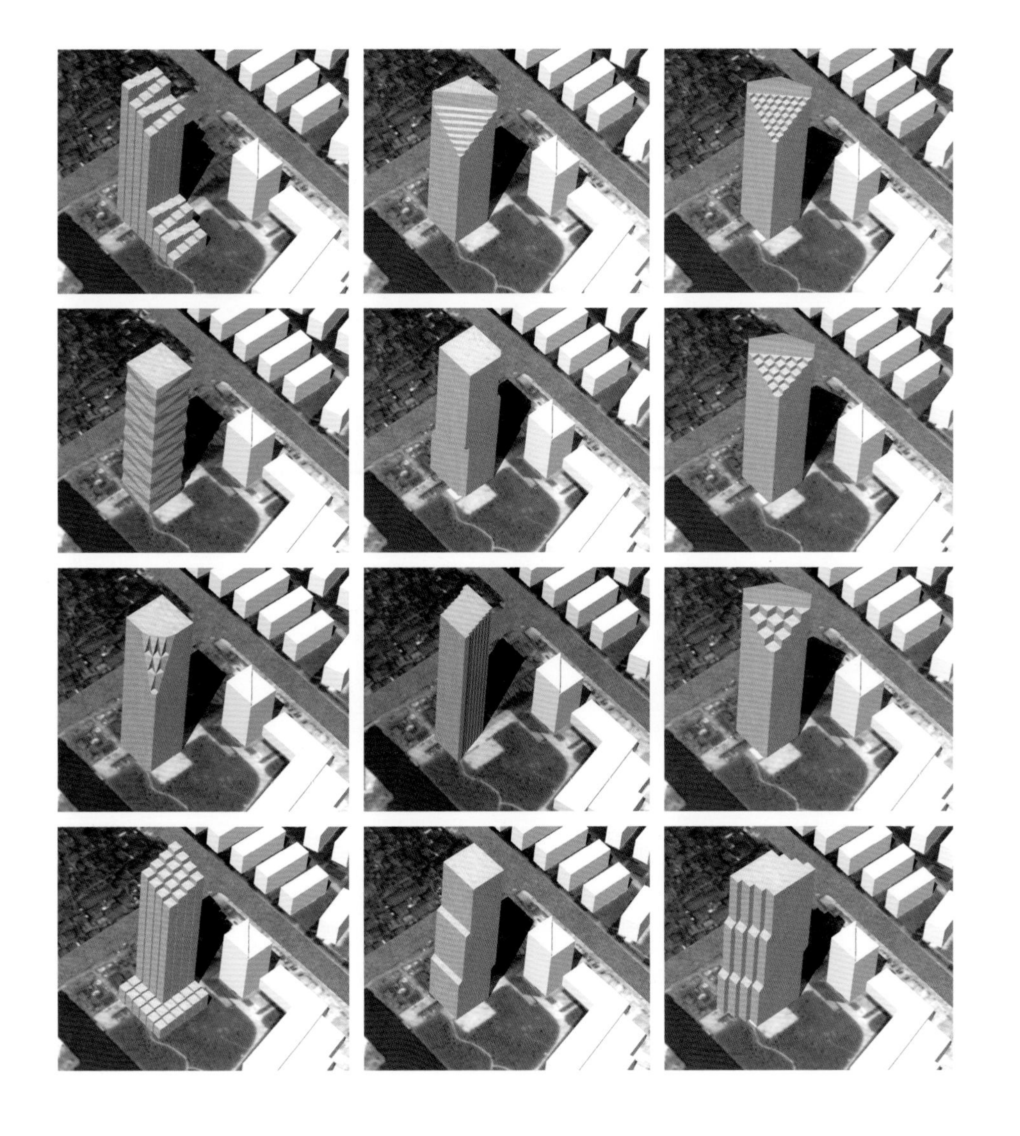

形体推敲

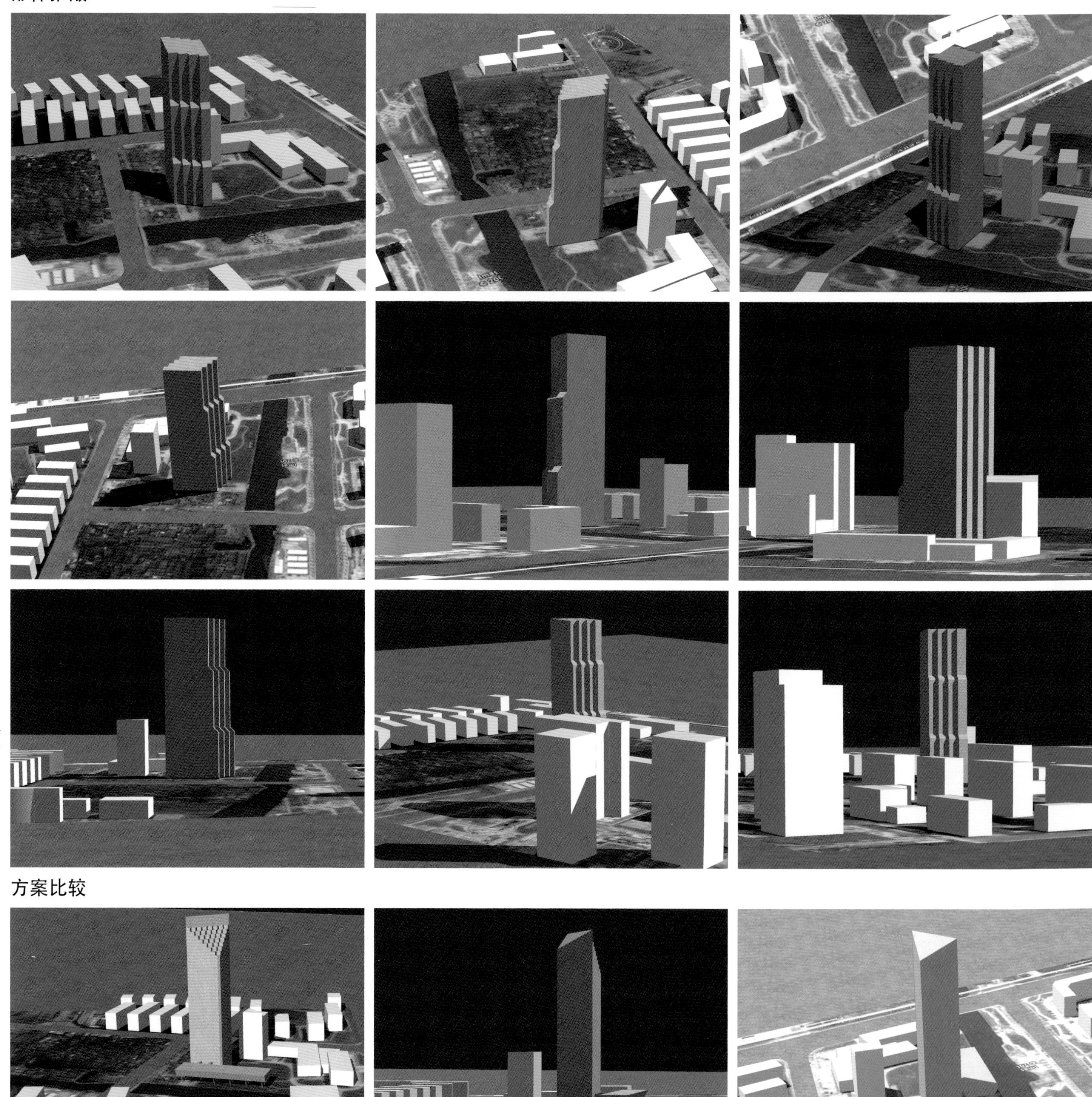

方案比较

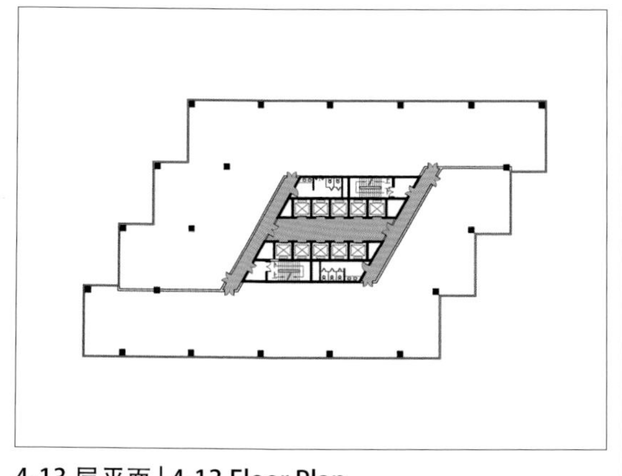

4-13 层平面 | 4-13 Floor Plan

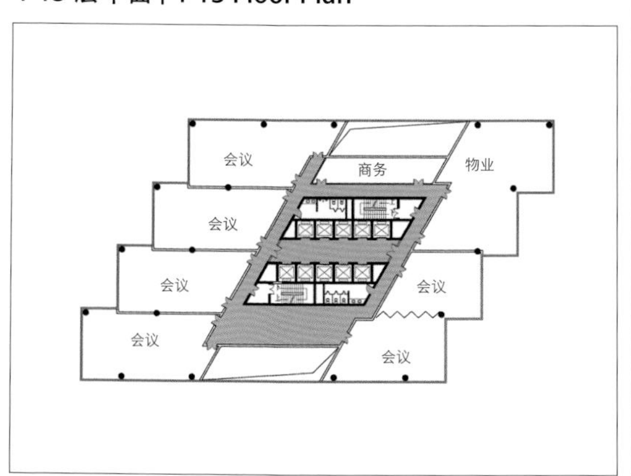

三层平面 | Third Floor Plan

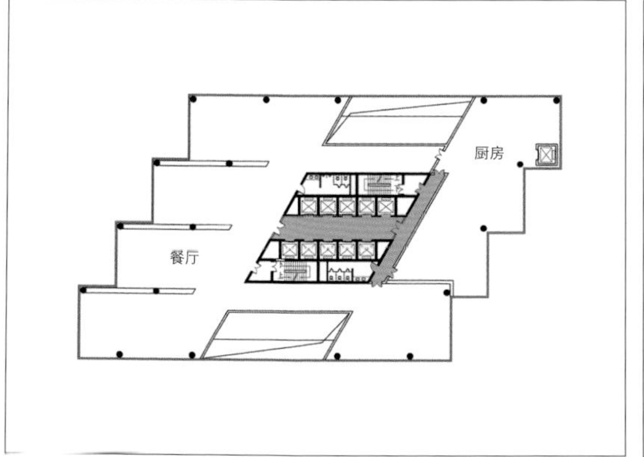

二层平面 | Second Floor Plan

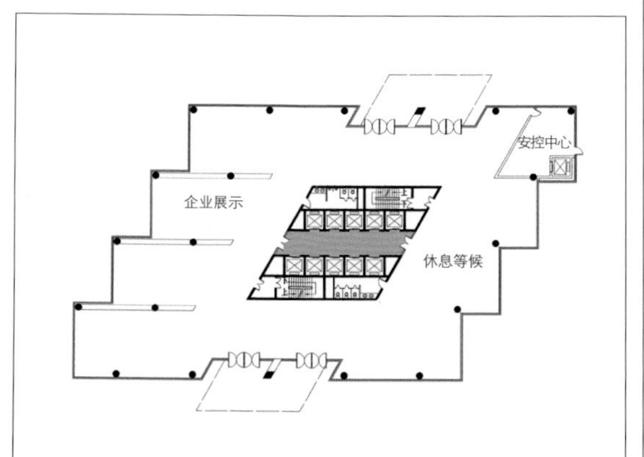

一层平面 | Ground Floor Plan

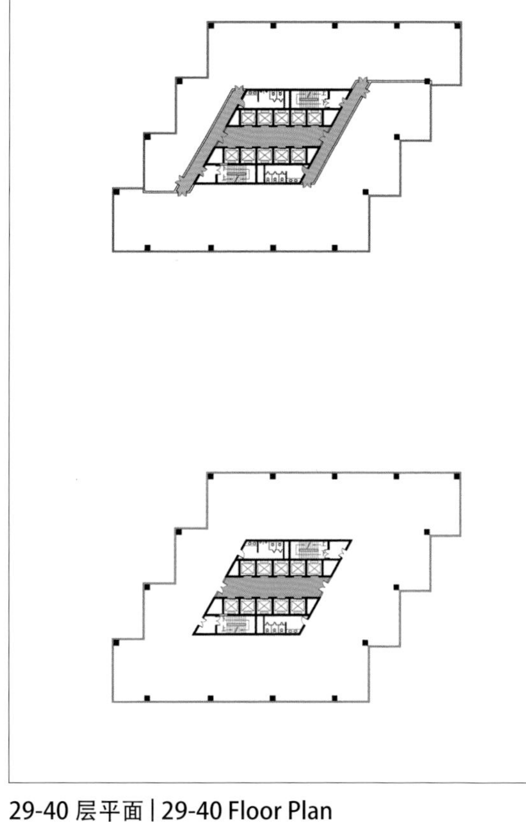

29-40 层平面 | 29-40 Floor Plan

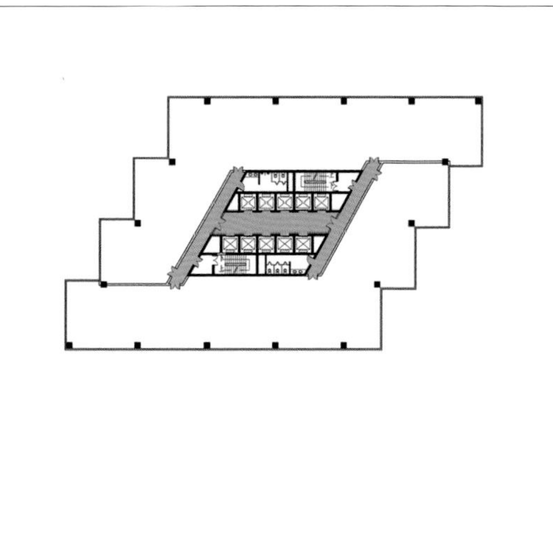

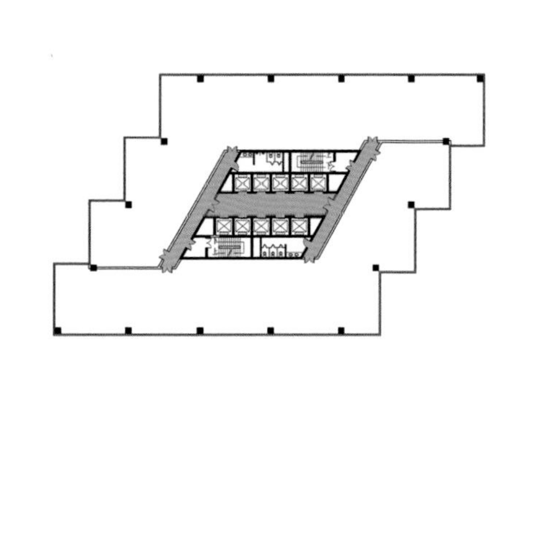

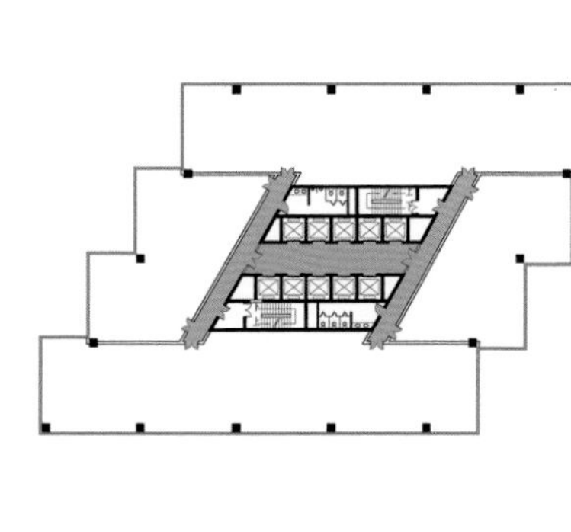

15-27 层平面 | 15-27 Floor Plan

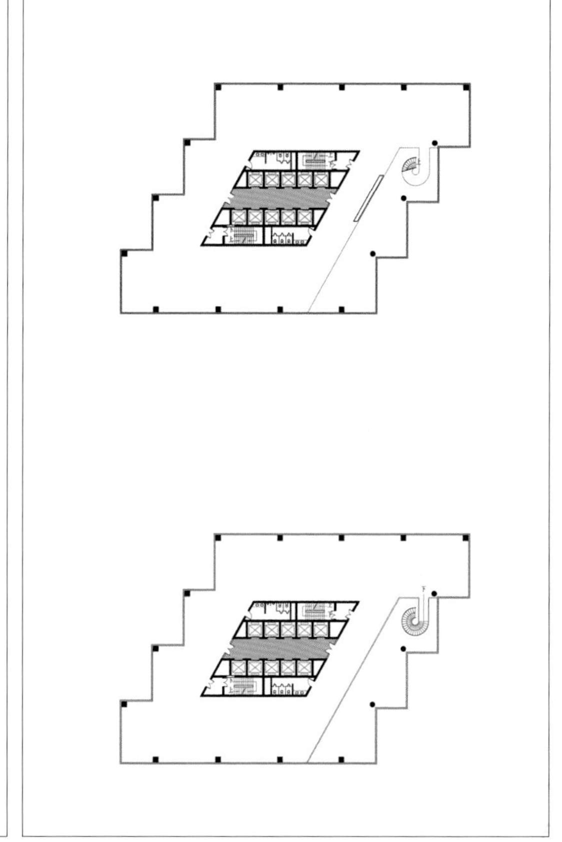

跃层平面 2 | Duplex Floor Plan

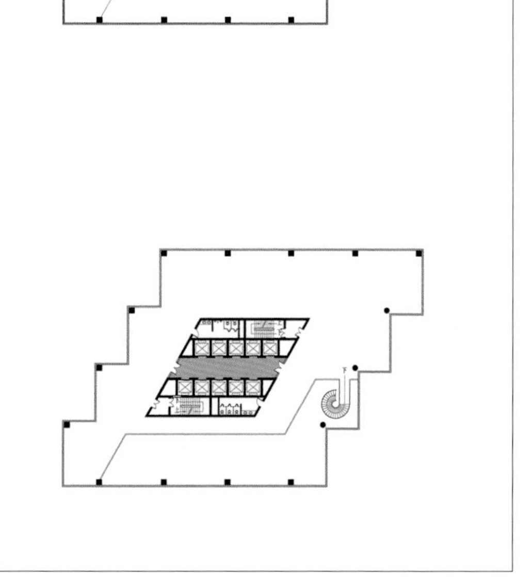

跃层平面 1 | Duplex Floor Plan

江苏省档案馆新馆

New Archives of Jiangsu Province

设 计 方 案　国际邀标 /2008
建筑师团队：高裕江、张弢、周敏、刘娜、陆信东

设计从形态上强化功能与形式的有机整合，并把握形体尺度，关注建筑的地域文化基因。形似或神似“孔明箱”的建筑形象突显象征中华民族智慧的文脉因素，体现生成建构的切合性。箱体形建筑界面肌理中嵌入十朝古都南京历朝建都的名称，东吴、东晋等，同时在其公共区域的虚体玻璃幕中“印刻”具有典型档案文化的印章，简列吴、周、黄、赵等中国十大姓氏，以诠释建筑的文化性、艺术性和时代性。

The design emphasizes the organic form-function relation through its morphology, the architectural yardstick and regional culture. The architecture is similar to "Kongming box" which highlights the chinese heritage and culture, and the pertinency of the construction. All the titles of ancient dynasties related to Nanjing, an ancient capital of ten dynasties, such as East Wu and East Jin are embodied in the texture of the architectural interface; in addition, the seals pertaining to typical local archives and local leading families such as Wu, Zhou, Huang and Zhao are stamped on the surface of glass in the public area so as to exhibit the architectural characters of cultural, artistic and time values.

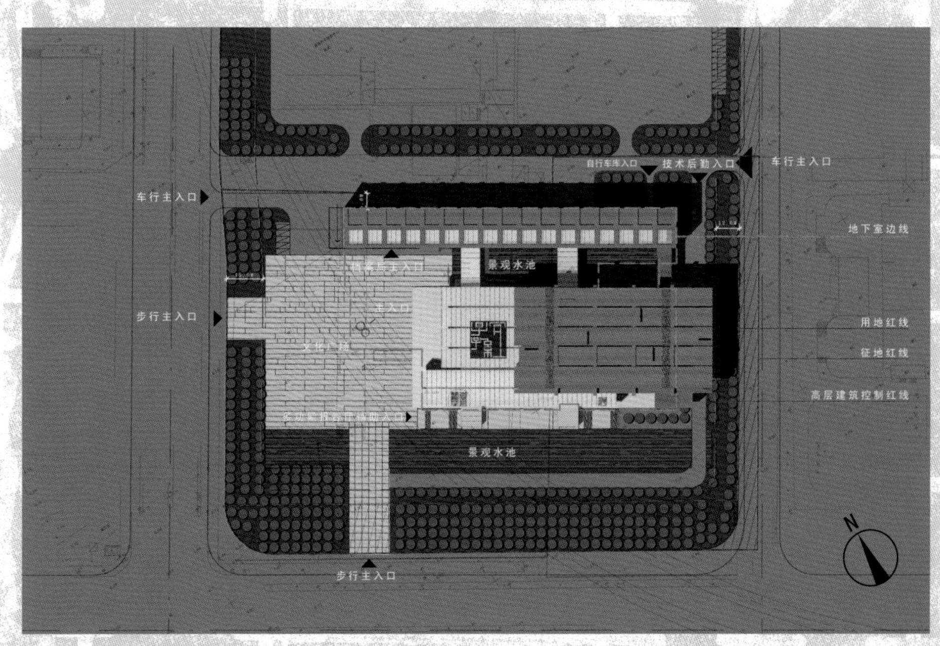

总平面图 | Site Plan

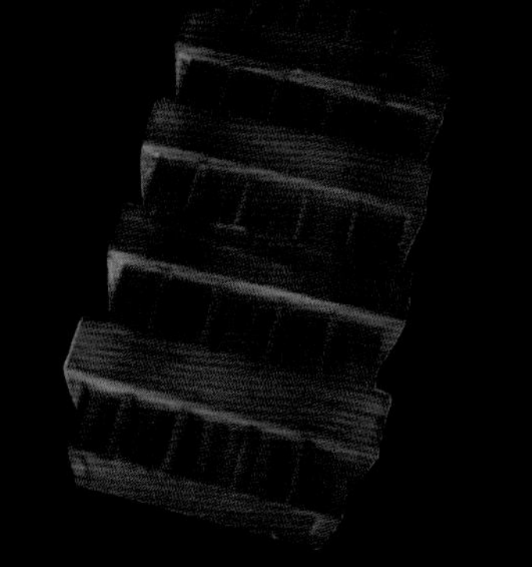

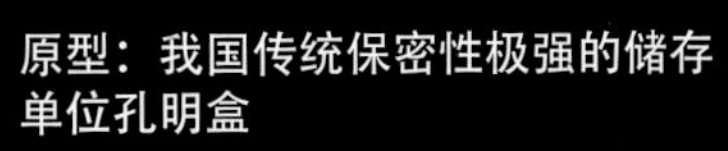

原型：我国传统保密性极强的储存单位孔明盒

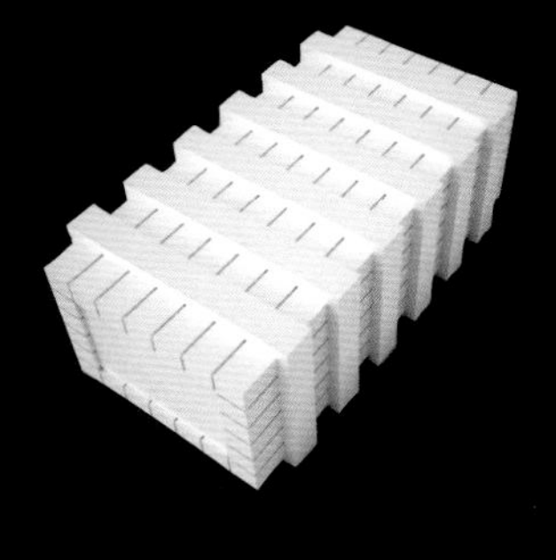

出发点

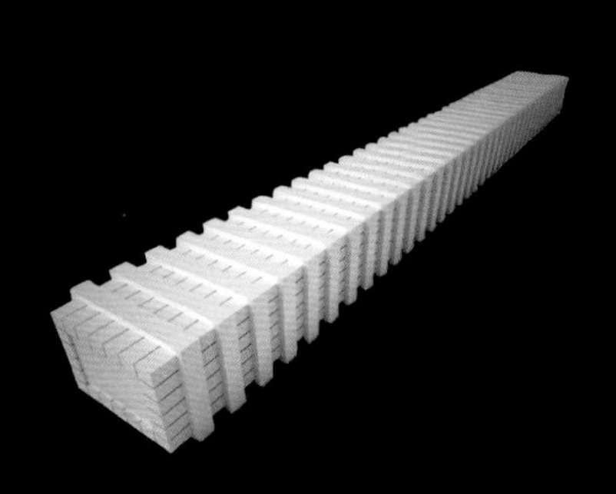

首尾相接为条状箱体

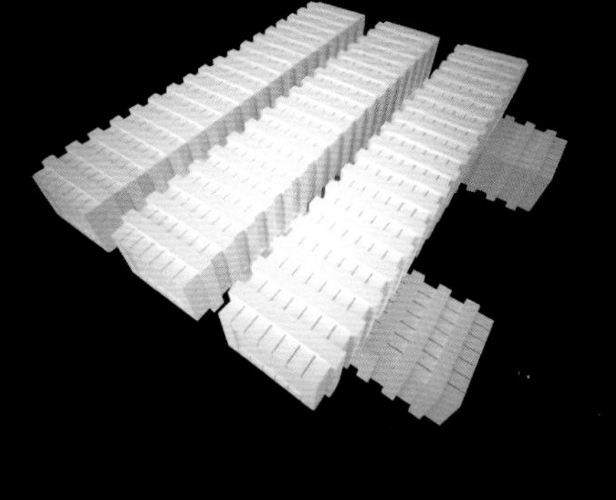

以条状箱体为单位初步搭接

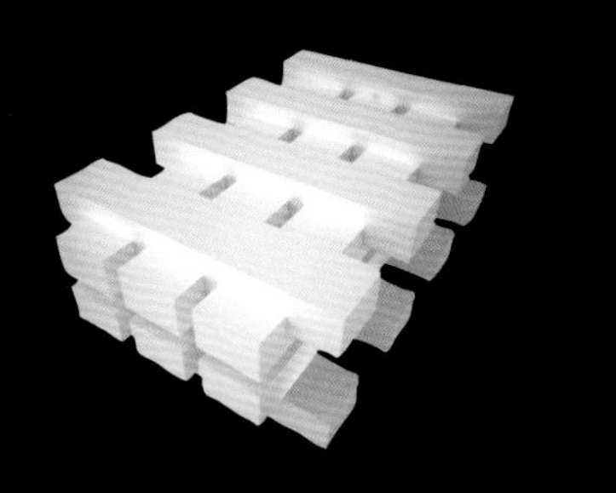

进一步叠加

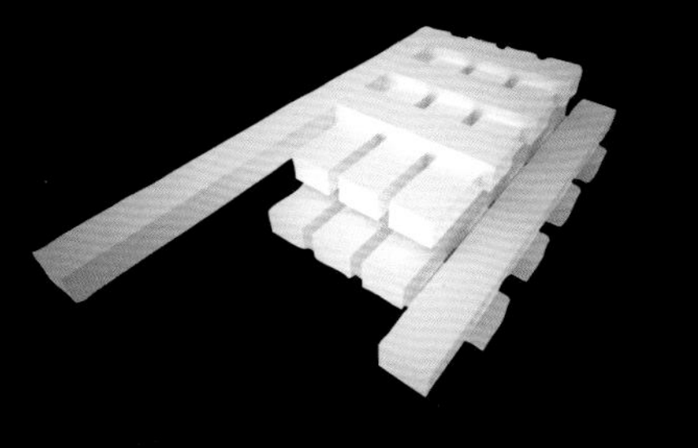

推敲形体关系

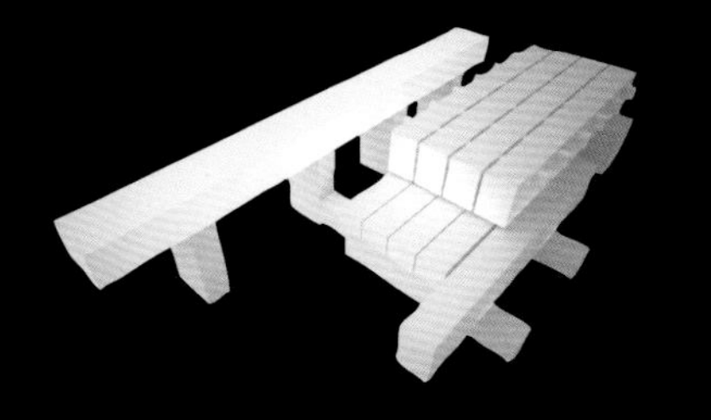

细化形体构成

成果

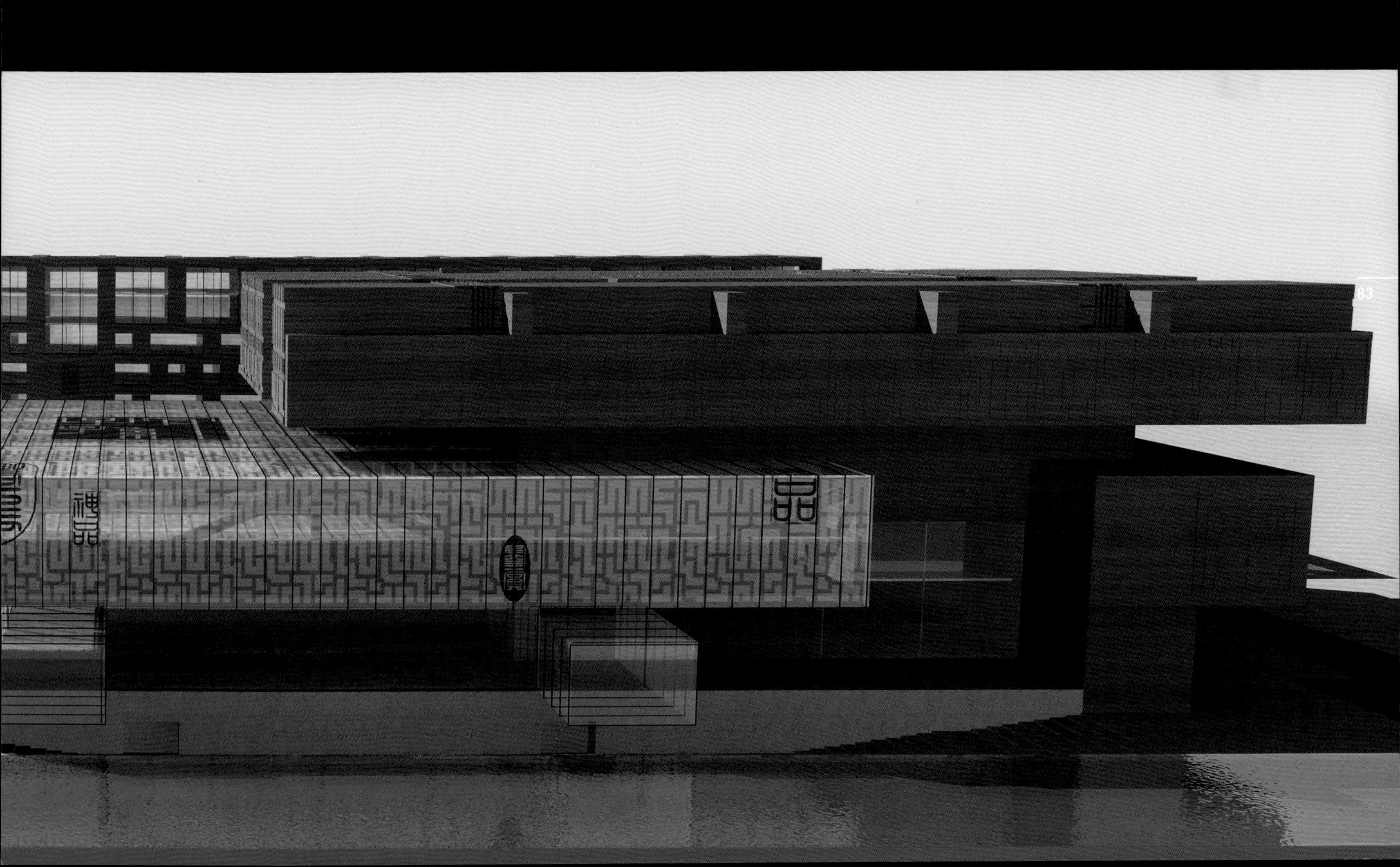

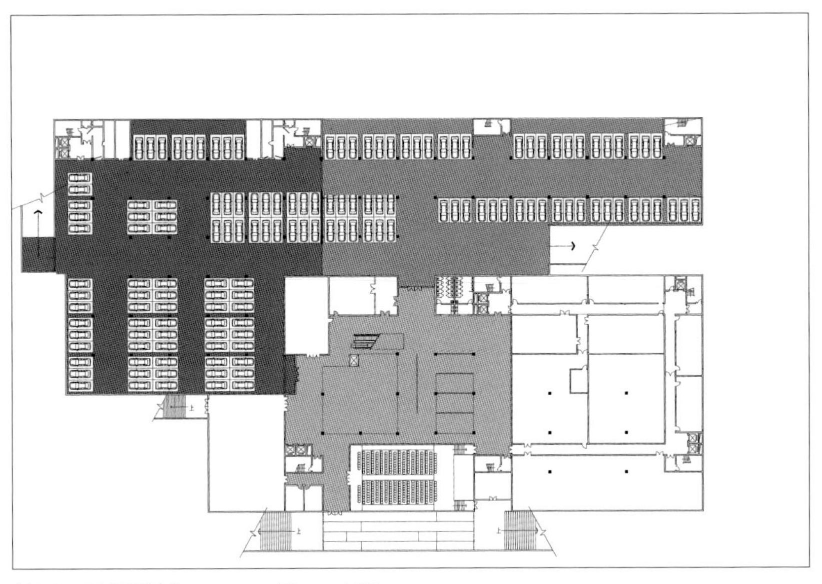

地下一层平面 | Basement Floor 1 Plan

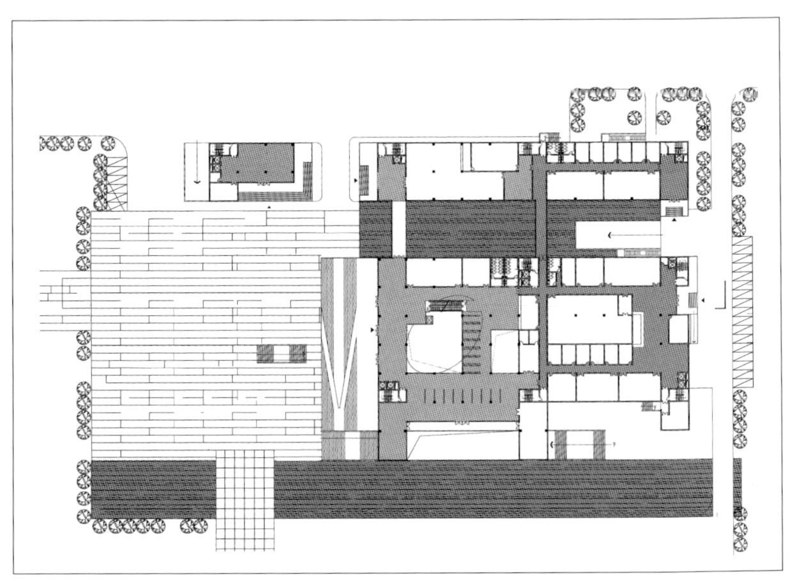

一层平面 | Ground Floor Plan

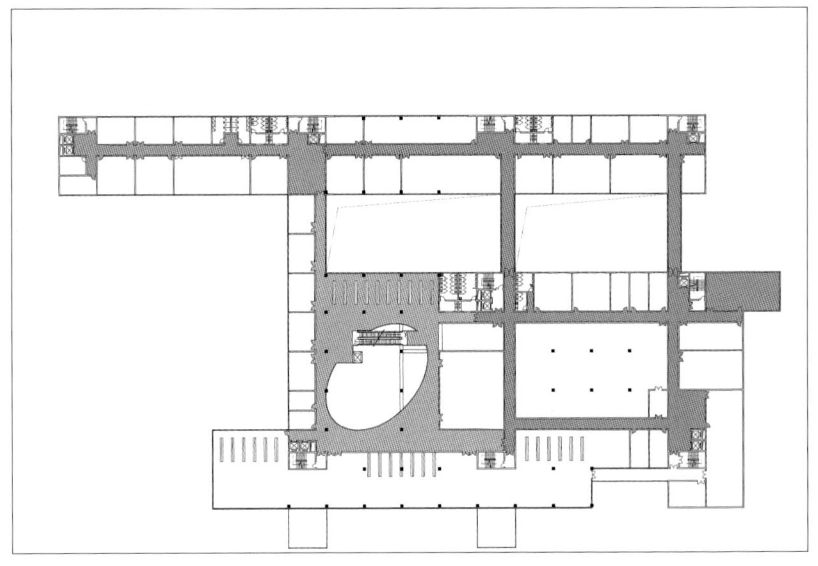

二层平面 | Second Floor Plan

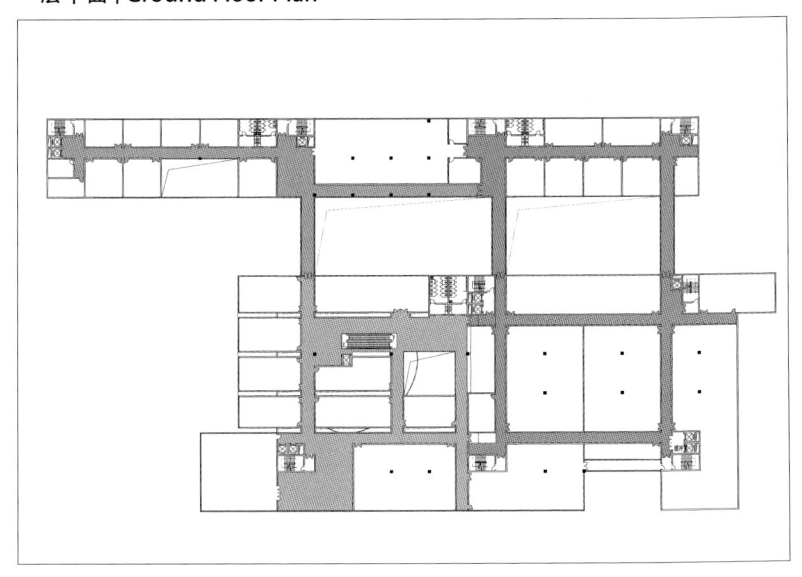

三层平面 | Third Floor Plan

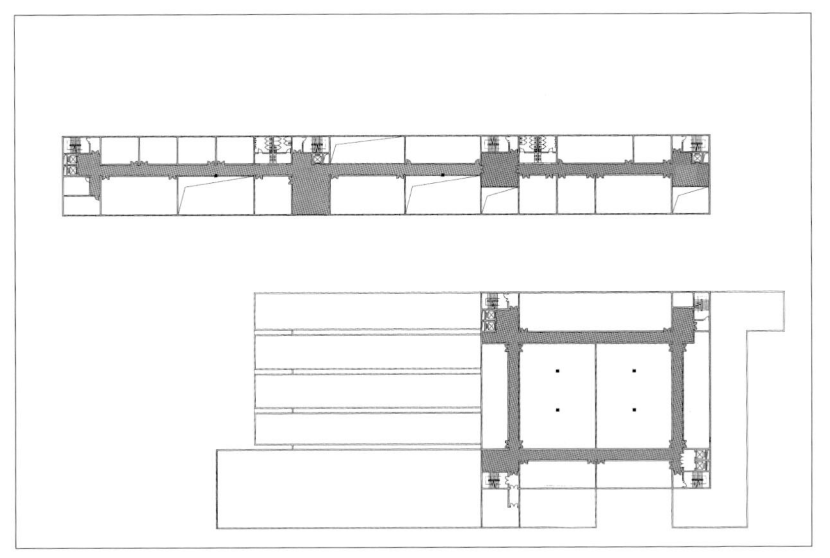

五层平面 | Fifth Floor Plan

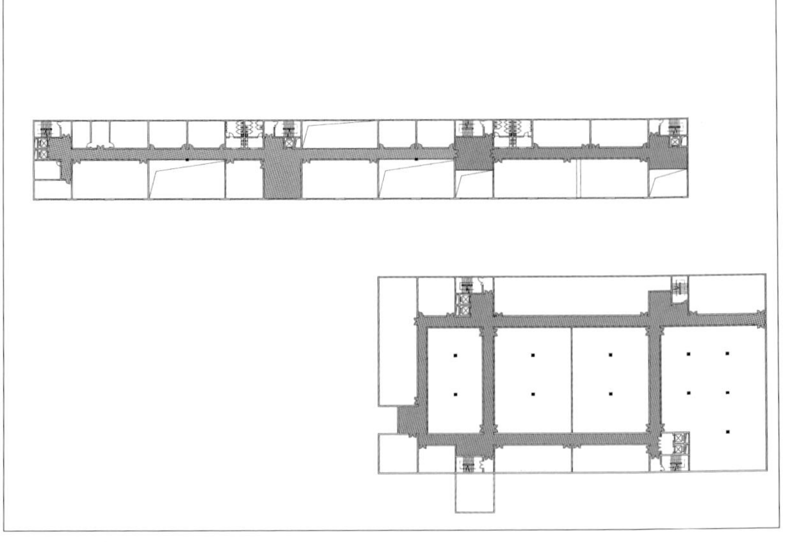

六层平面 | Sixth Floor Plan

江苏省省级机关河西办公区

Office Buildings of Province-Level Organs in Jiangsu Province

概 念 规 划　国际竞标规划方案 /2008
建筑师团队：高裕江、张弢、陆卫兵、刘娜、周敏、陆信东

十个厅局级行政单位，以均等的形式组成五重内院。主体办公南北布置并居上方，辅助办公位居东西两侧下部，形成“L 形”的平面体系，内外有机融合。设计通过多重传统院落的空间模式，使建筑群整构合一，强化生态节能的同时，承延“传统大院”的归属感，突显内敛的空间文化特色，从而保障了整体办公环境的上佳品质。

Ten bureau-level administrative units equally constitute five courtyards. The main official area is north-south oriented in the upper position while the secondary official area is east-west oriented at the bottom , to form L-shaped spatial complex and the organic fused space of inside and outside.The design adopts the spatial model of traditional multi-courtyard in order to reconstruct the building complex and highlight the ecological energy saving. It not only inherits the feeling of belonging related to "traditional courtyards" , but also pinpoints the internal spatial cultural characteristics, to improve the spatial quality of the whole office building.

推敲过程

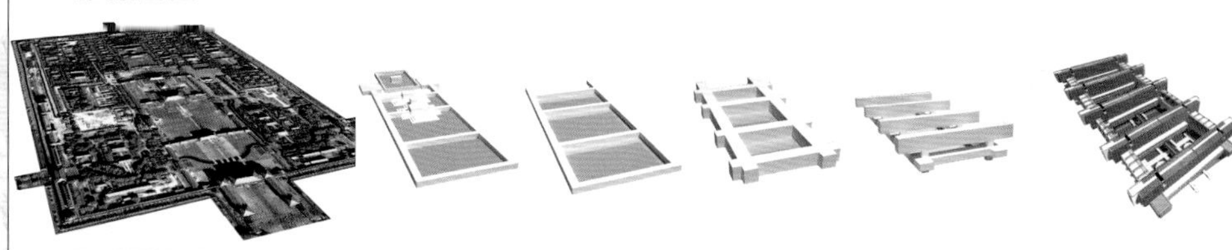

模型照片

规划分析

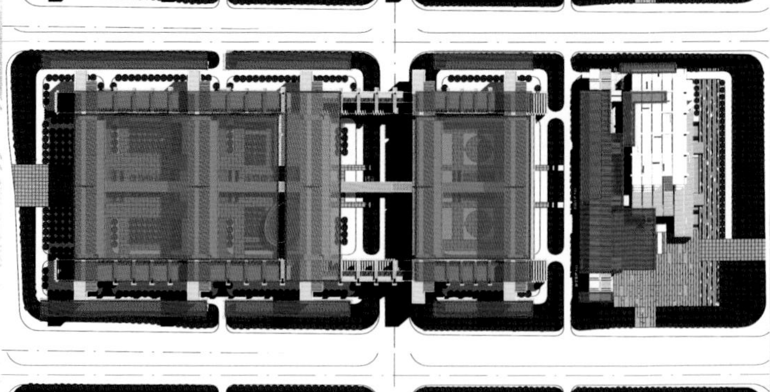

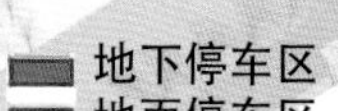

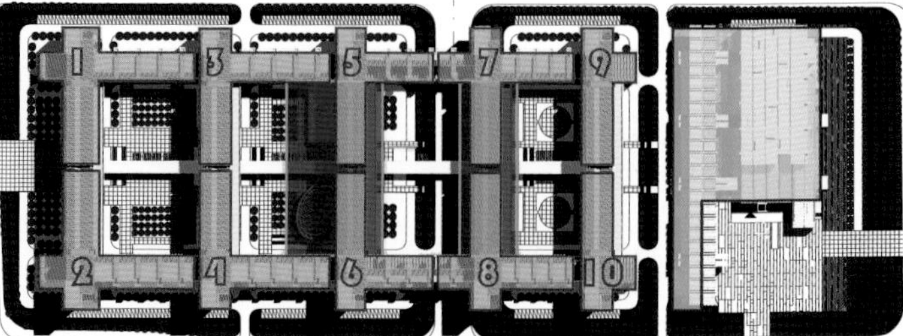

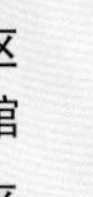

总平面图 | Site Plan

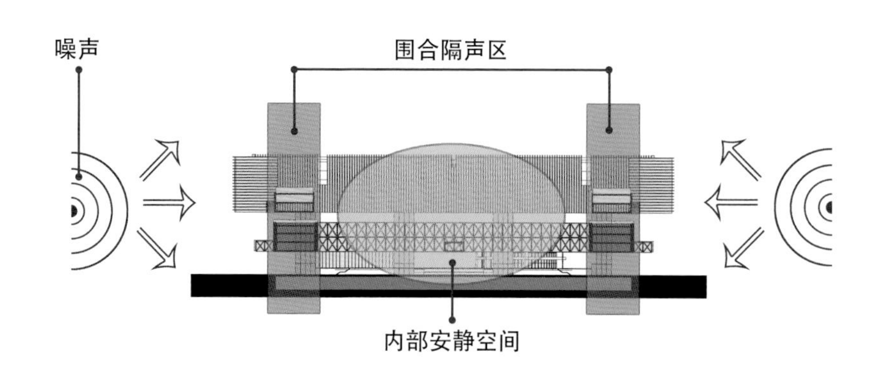
噪声
围合隔声区
内部安静空间

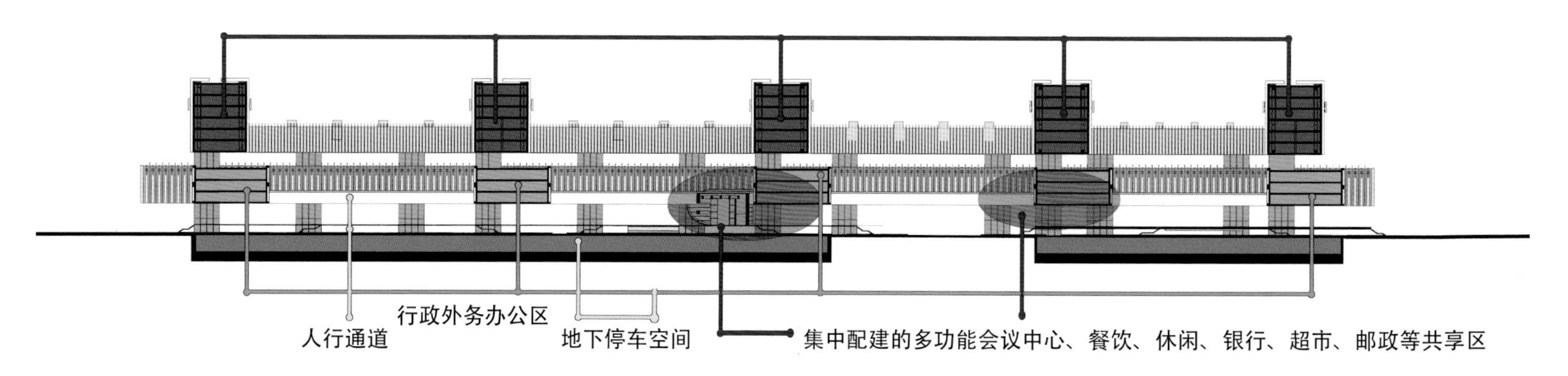
行政外务办公区
人行通道
地下停车空间
集中配建的多功能会议中心、餐饮、休闲、银行、超市、邮政等共享区

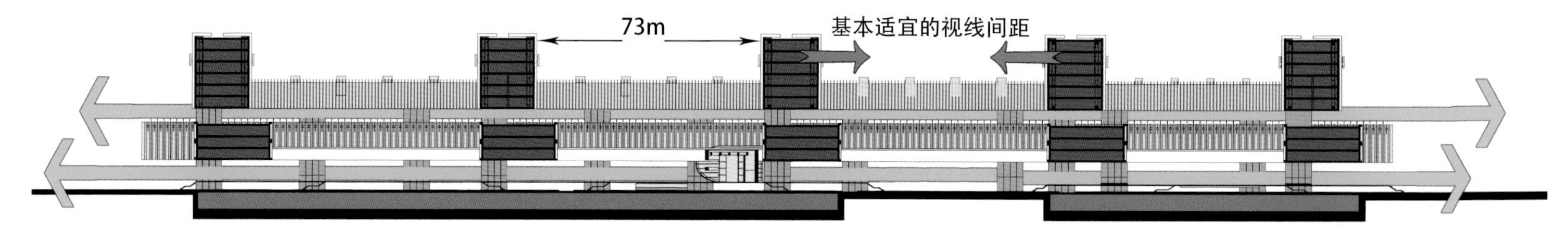
73m
基本适宜的视线间距

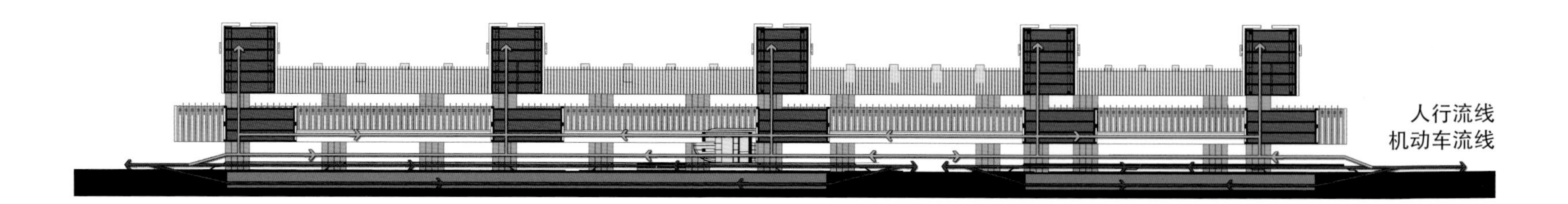
人行流线
机动车流线

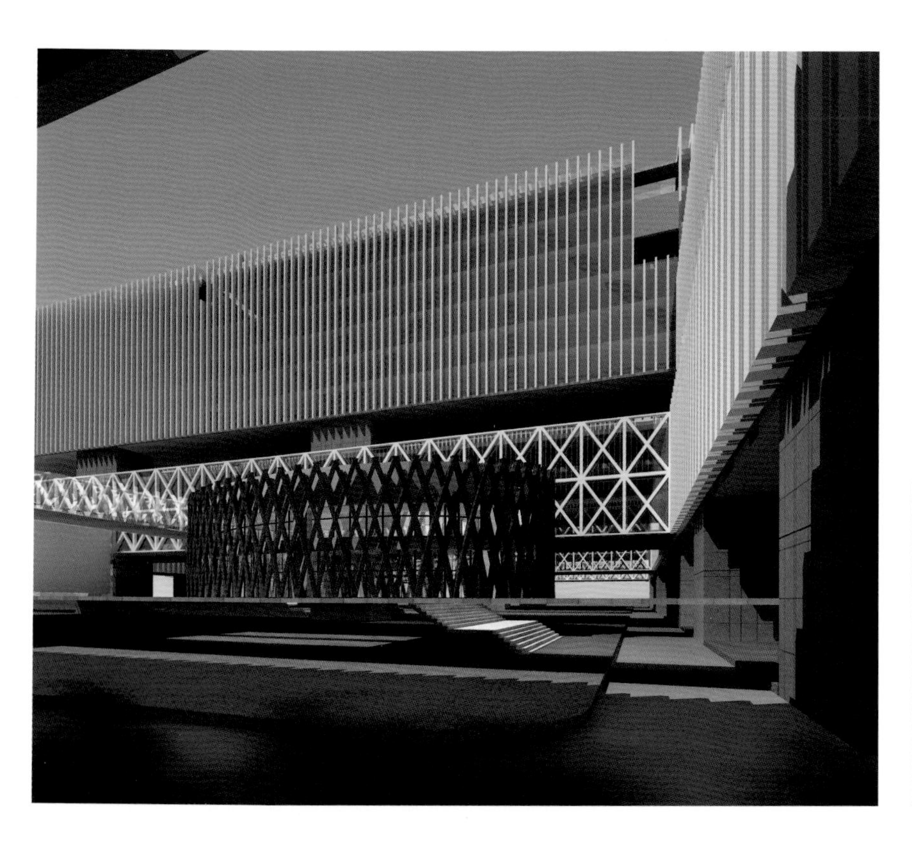

宁波鄞州云龙镇商业综合街区概念方案

Conceptual Design of Complex Commercial Block in Yunlong Town, Yinzhou, Ningbo

概念规划（实施） 2008
方　　案　　图：高裕江、赵甜甜、顾月明
施　　工　　图：宁波鄞州设计研究院

设计通过商业街、巷、广场、内院的有机整构，形成极其丰富的商业空间氛围。在有利于不同业态的分布和设置、提升商业价值的同时，传递出传统村落、街巷的空间景境。建筑群体的形象风格、街巷空间尺度，以及建筑材质肌理，都源于传统民居和街巷的传承、延展和变异。

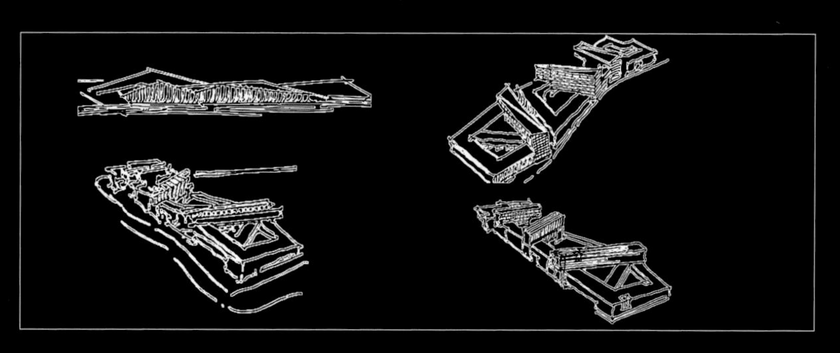

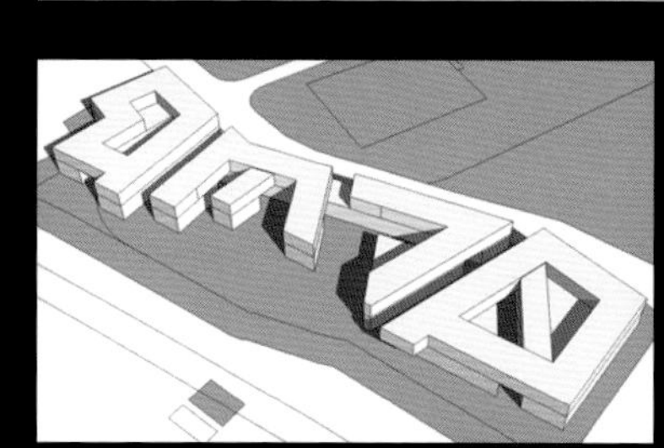

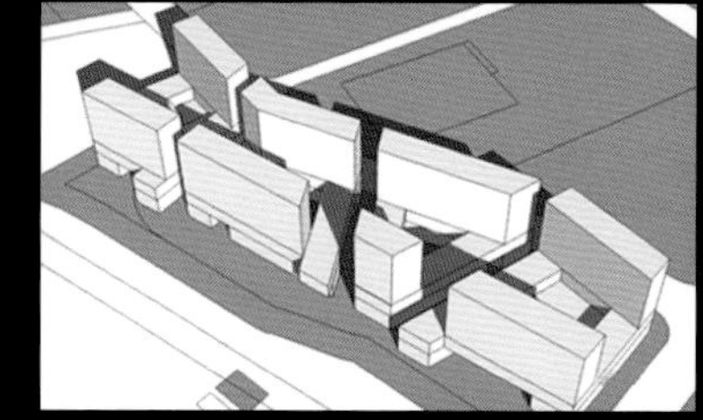

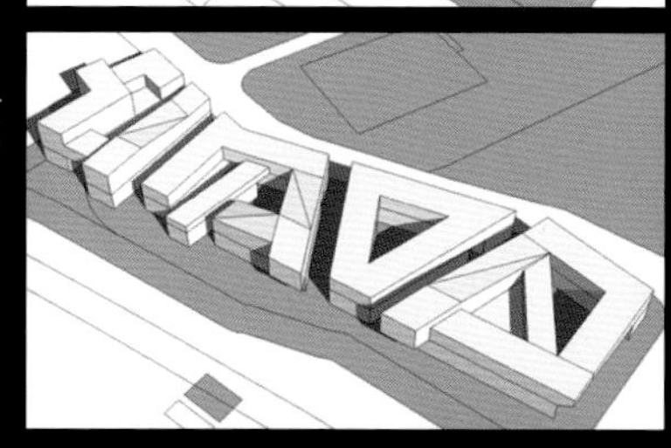

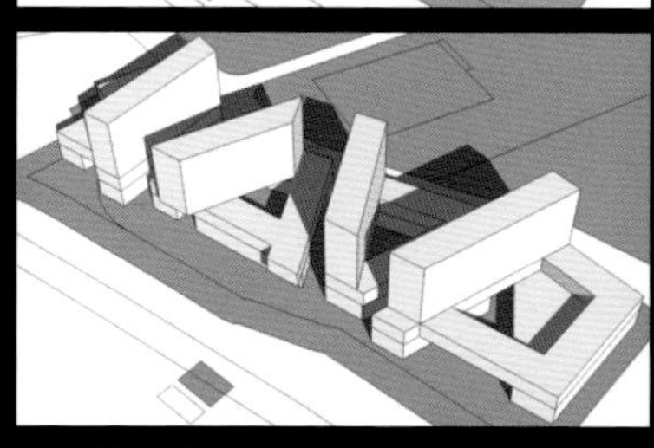

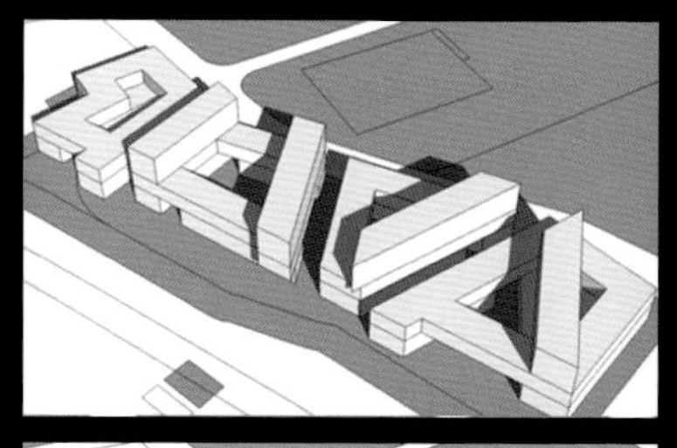

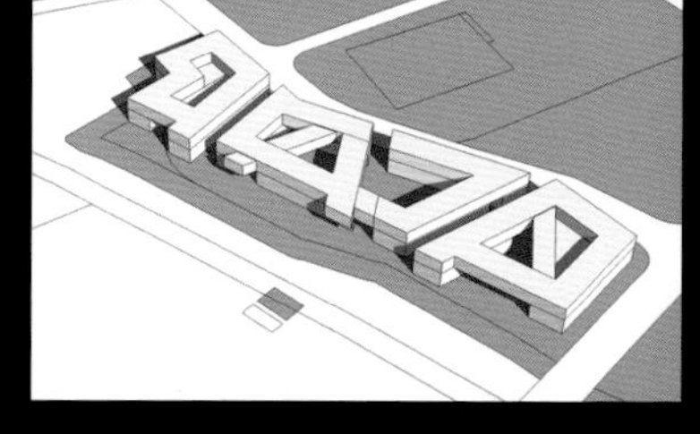

形体演化

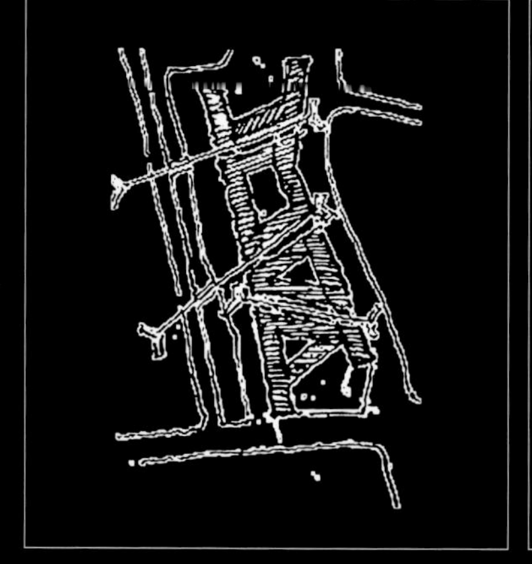

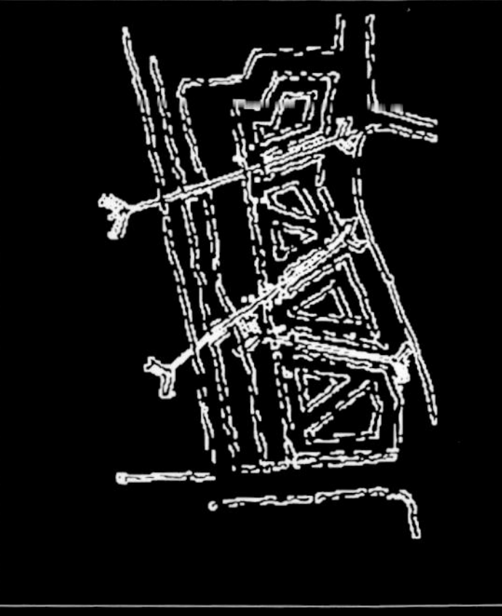

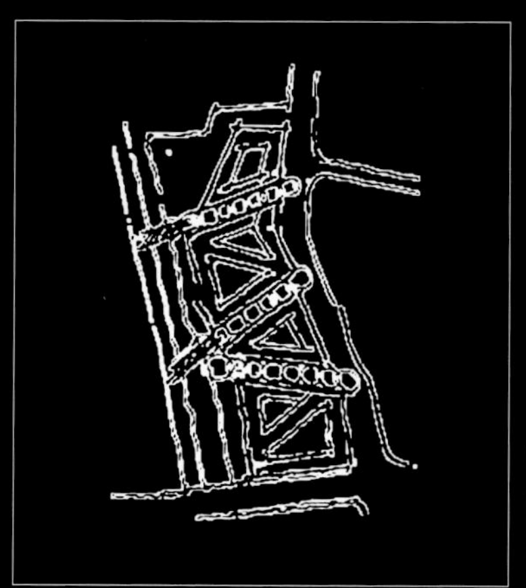

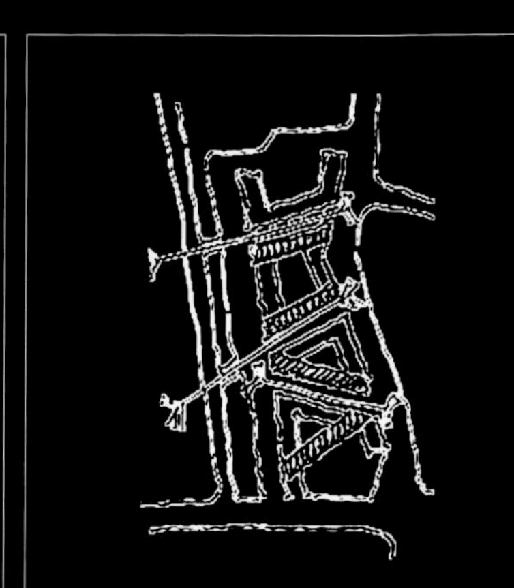

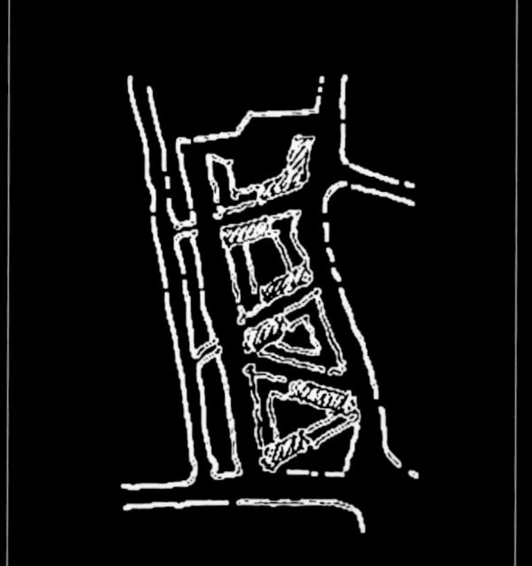

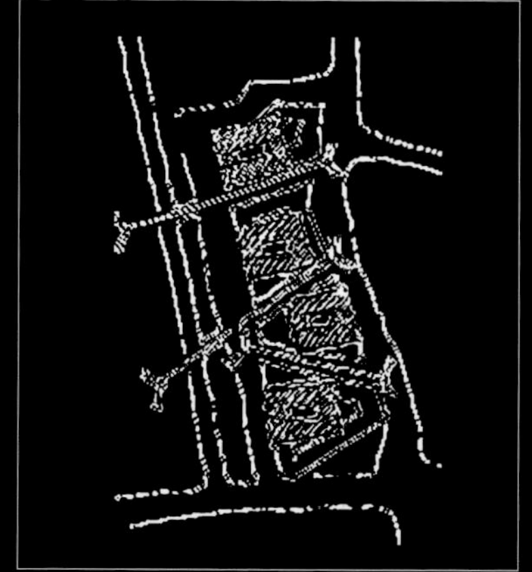

草图 | Sketch

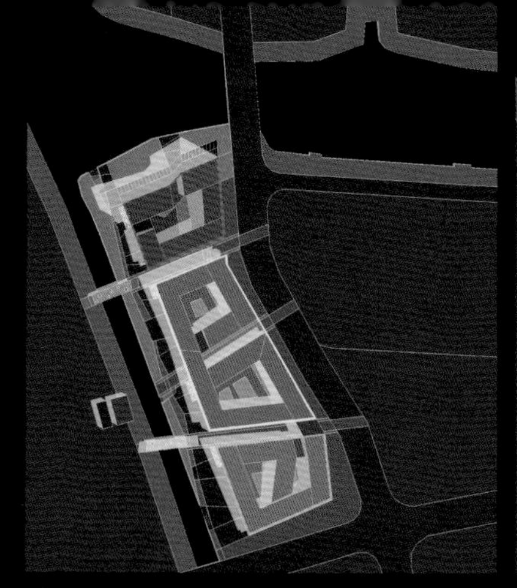

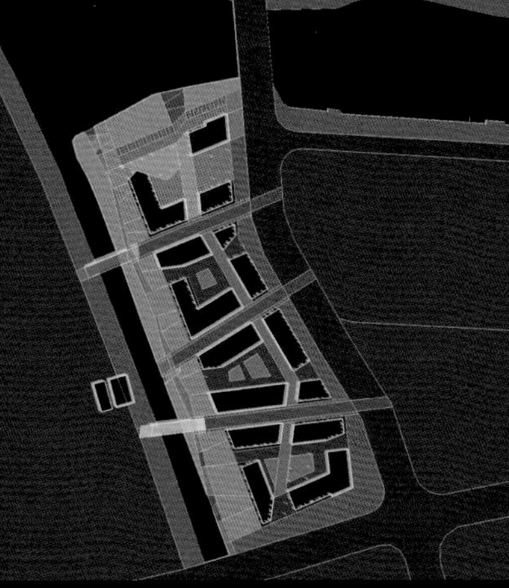

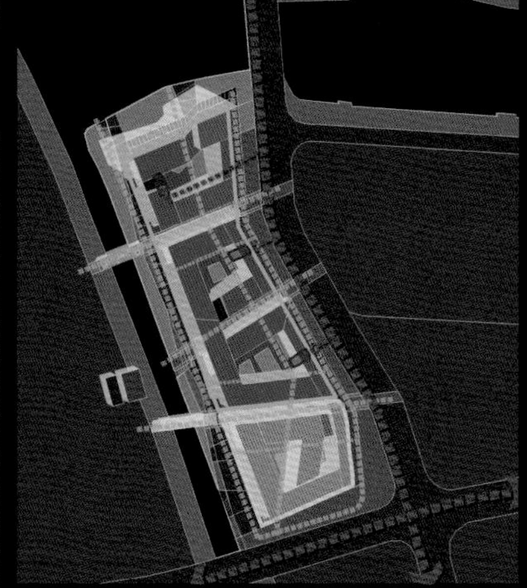

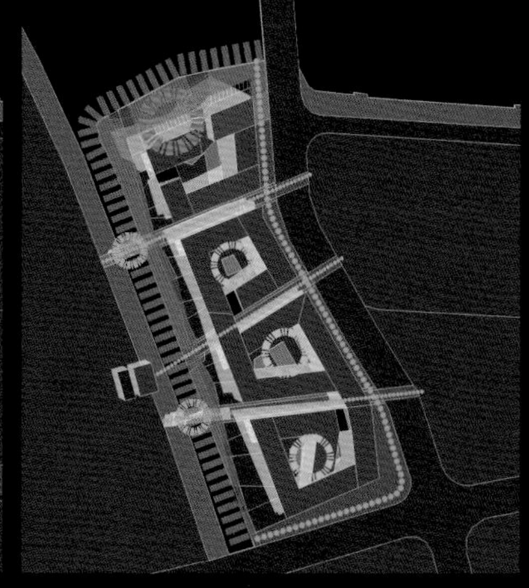

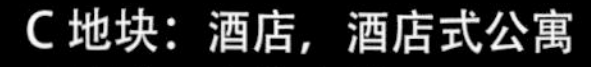

C 地块：酒店，酒店式公寓
B 地块：出售商铺
A 地块：回迁商铺 地下商场

城市缓冲空间
商业空间
滨水景观休闲空间
滨水生态带
酒店外部空间

地下车库
地下商场
地下车库入口
机动车流线
人行流线

近地水域
滨河绿化带
沿街绿化带
内院绿化
视线通廊

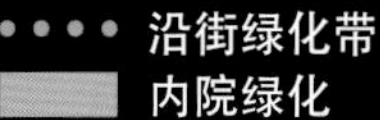

滨河景观节点

景观步行桥

The design, through the organic organization of commercial streets, alleys, squares and courtyards, tends to form commercial spaces with affluent gradations. It not only benefits the layout of different functions, but also improves the commercial value. In addition, it implies the scenery of traditional villages and streets. The architectural style, street scale and architectural material texture are all derived from inheriting, extension and variation of the traditional vernacular dwellings and streets.

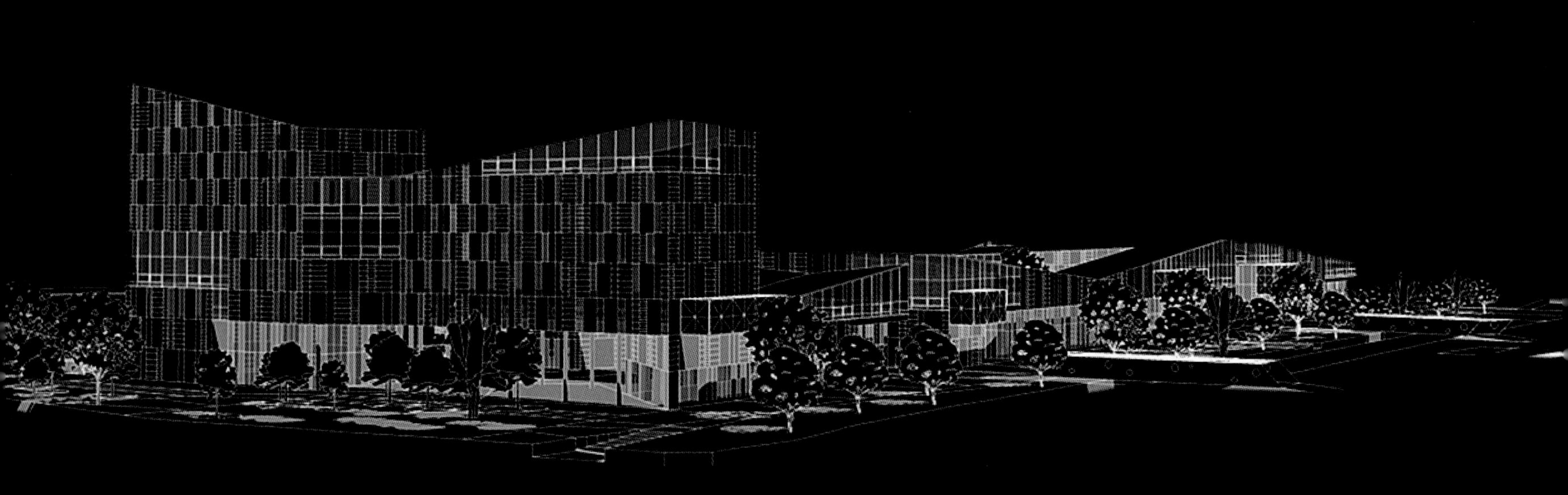

南京博物院扩改建工程

Renovation and Extension Projects of Nanjing Museum

概 念 方 案　国际竞标 /2008
建筑师团队：高裕江、周敏、刘娜、陆卫兵、陆信东

设计采用“城院整构”的格局，强化延续原历史馆的中轴线，在博物院的前部形成源自中华门瓮城型入口广场，呼应中山门，提升博物院的空间形象，屏障不利的外部环境，营造出博物院惯有的庄重、优雅、宁静，而又具历史性、艺术性、文化性氛围。设计充分挖掘现有用地的各种潜力，完善功能架构，同时预留发展空间。运用积极的、扩建、保护整合的策略，改善历史馆老大殿与主庭院空间关系，完善空间序列，强化西侧及其他相关空间轴线。中心建筑——博物院中央大堂之形式源于“传统中国土金字塔”形制，通过转换、变异、叠合形成一种全新的建筑体形象。

Adopting the form of “reconstruction of courtyard”, the design highlights and extends the axis of the original history museum and forms the entrance square derived from Barbican of Zhonghua Gate. The arrangement reflects Zhongshan Gate, improves the spatial visualisation of the museum and gets rid of the unfriendly external environment and forms the solemn, elegant and peaceful feelings and historical, artistic and cultural environment. making use of the original art gallery, the design takes full advantage of all the potentialities of the premise and improves its functional framework. meanwhile, a “niche” is reserved for the further development.The design adopts active strategies of extension and reservation in order to improve the spatial relation between the main hall of history museum and the main courtyard, Improve the spatial sequences and highlight the west and other related axes .The form of the central architecture—main hall of the museum is derived from the shape of traditional Chinese Pyramid”. The brand new architectural morphology is achieved by spatial transformation, variation and overlap. It expresses the senses of permeability, lightness and creativity.

形态演变

土金字塔形态演变

传统庑殿顶形态演变

聚宝盆

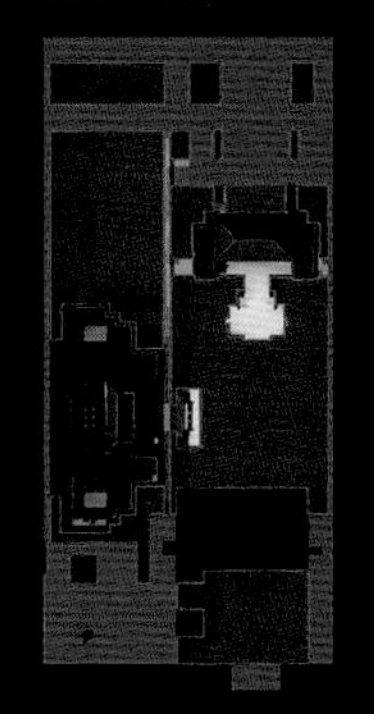

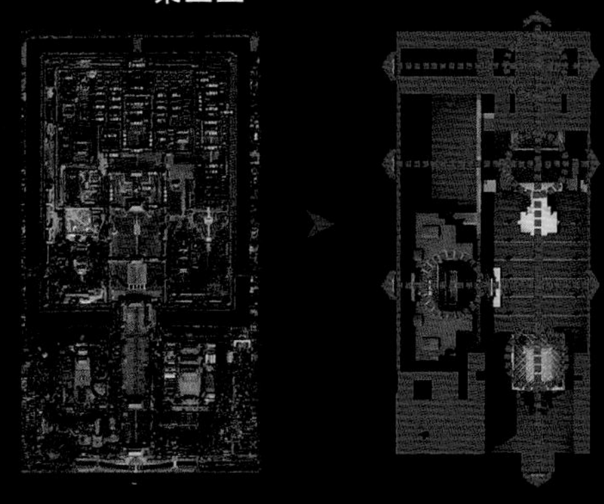

古代城院形态演变

平面图 | Floor Plan

方案比较

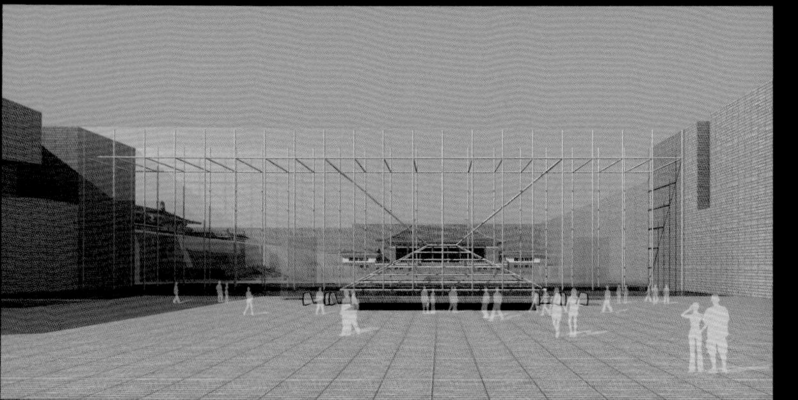

宁波云龙文化中心

Ningbo Yunlong Cultural Center

实 施 工 程　2008—2012
建筑师团队：高裕江、顾月明、陆卫兵、周敏、
刘娜、郑颖生、史国雷、饶峥、
丁思璐、赵甜甜

鉴于近山滨河及“L 形”道路转角用地的现状，设计将影剧院和体育馆两组楔形体，依据城市空间肌理相向置于基地中，形成动态、对话之势。在功能与形态组构中，运用“加减法”的设计手法，建筑形象准确而真实地显示内部空间特征。在建筑界面营建上，吸纳信息时代特有的文化性符号，以及地域建筑文化元素的承延运用，从而展示出文化性、时代性和艺术性。

As the site is near mountains and rivers, and connects with L-shaped road corner, two sets of wedge-shaped spaces (theatre and gymnasium) toward each other according to the urban texture, and form the trending of movement and dialogue with each other. In the aspect of form-function relation, the design applies "adding and subtracting strategies"; thus, the architectural images correctly reflect the internal spatial characteristics. In the aspect of architectural interface construction, the cultural characteristics in the Information Era are absorbed in the design and regional architectural cultural elements are inherited, in order to exhibit the architectural characters of cultural, time and artistic values.

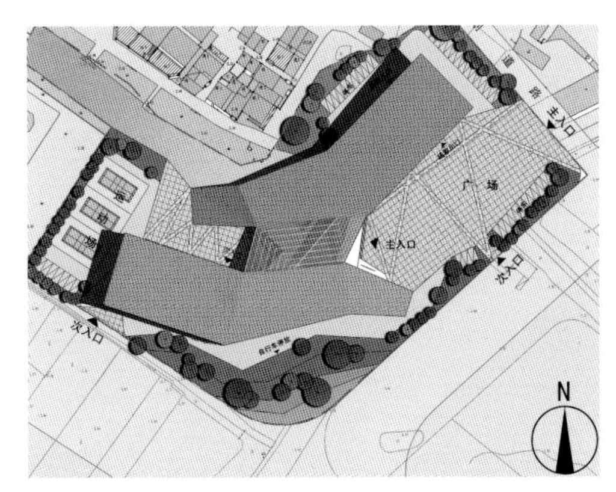

总平面图 | Site Plan

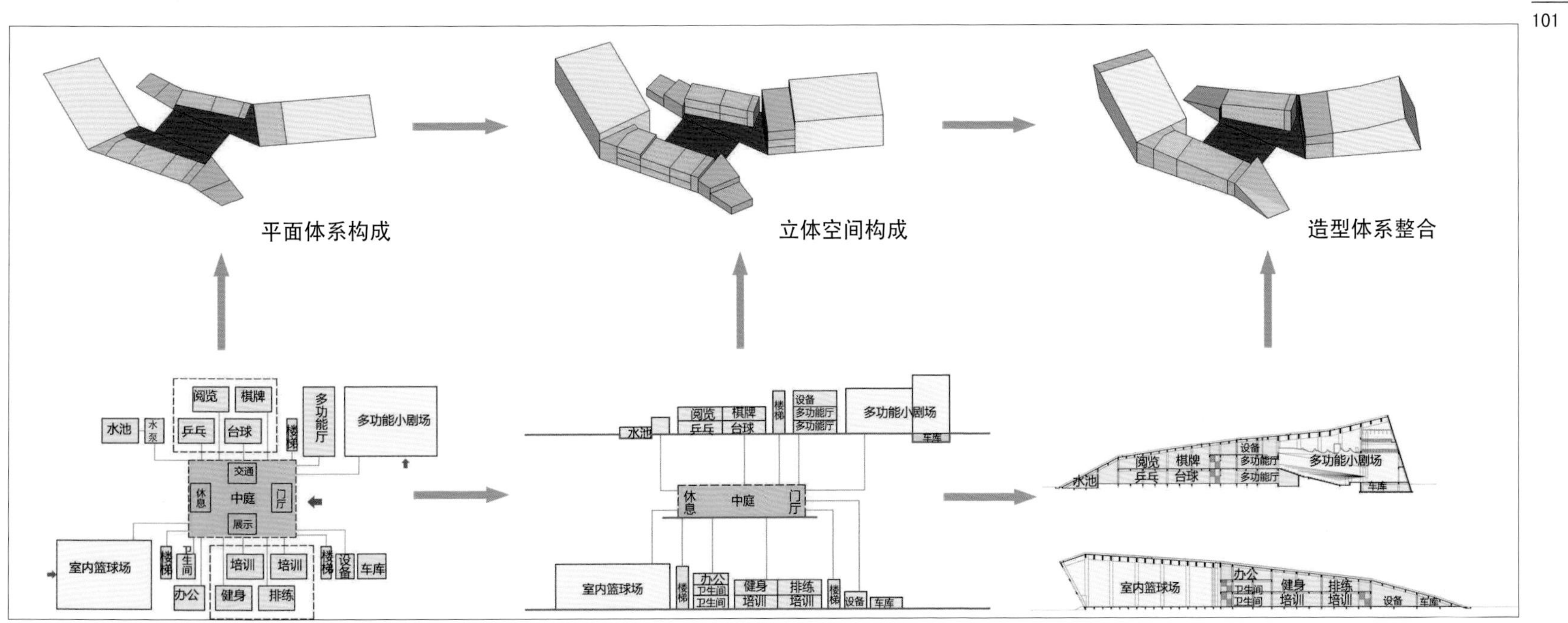
平面体系构成
立体空间构成
造型体系整合
阅览
棋牌
水池
水泵
乒乓
台球
楼梯
多功能厅
多功能小剧场
交通
休息
中庭
门厅
展示
室内篮球场
楼梯
卫生间
培训
培训
楼梯
设备
车库
办公
健身
排练
水池
阅览
棋牌
乒乓
台球
楼梯
设备
多功能厅
多功能厅
多功能小剧场
车库
休息
中庭
门厅
室内篮球场
楼梯
办公
卫生间
卫生间
健身
培训
排练
培训
楼梯
设备
车库
水池
阅览
棋牌
乒乓
台球
设备
多功能厅
多功能厅
多功能小剧场
车库
室内篮球场
办公
卫生间
卫生间
健身
培训
排练
培训
设备
车库

能分析

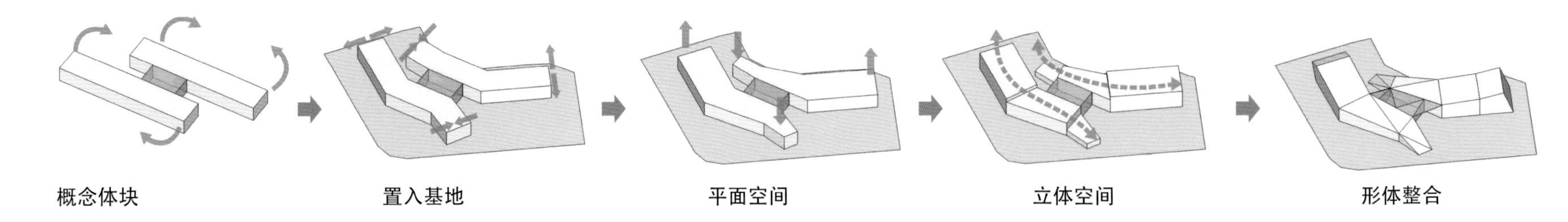

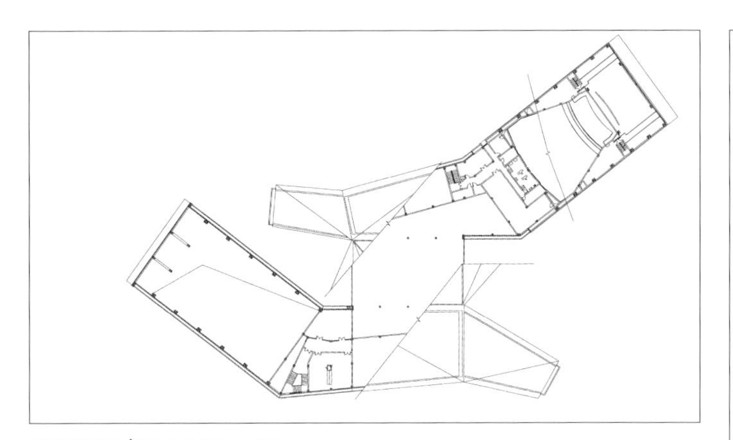

三层平面 | Third Floor Plan

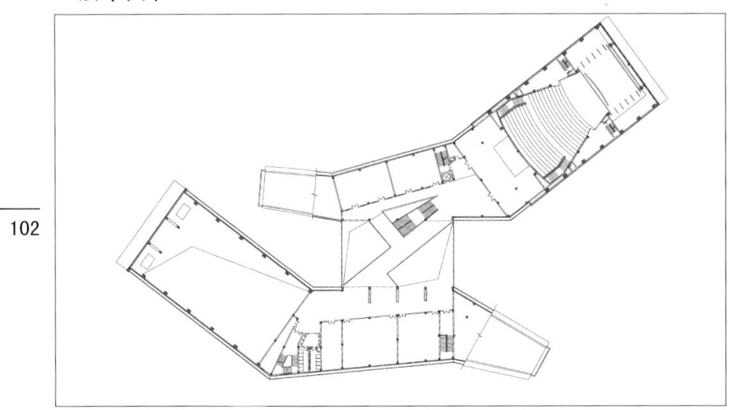

二层平面 | Second Floor Plan

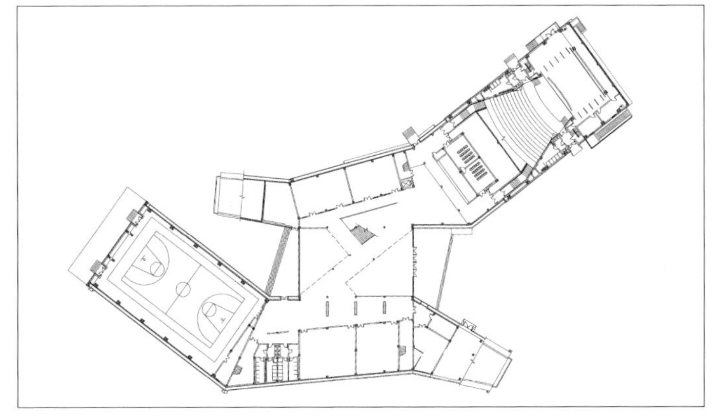

一层平面 | Ground Floor Plan

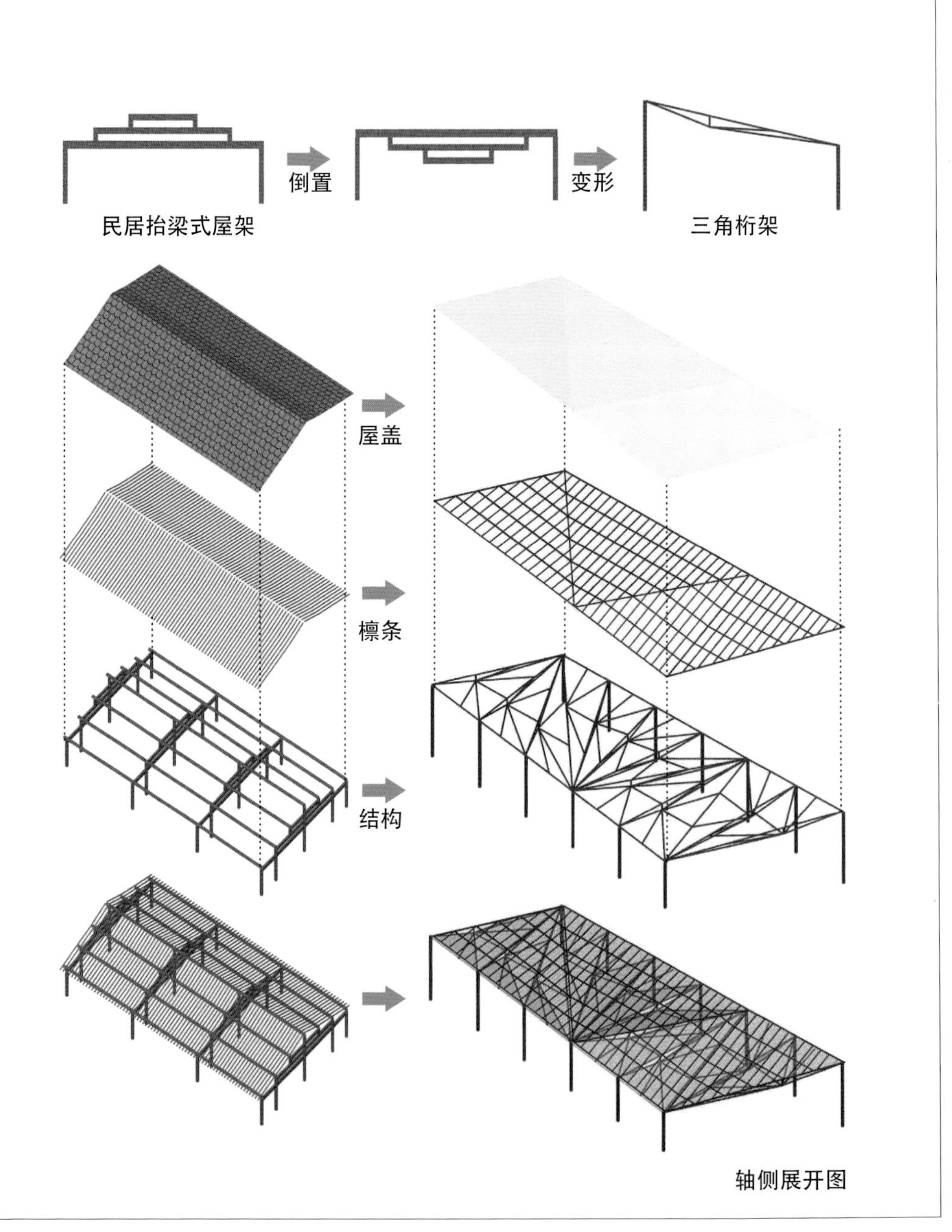

桁架概念生成

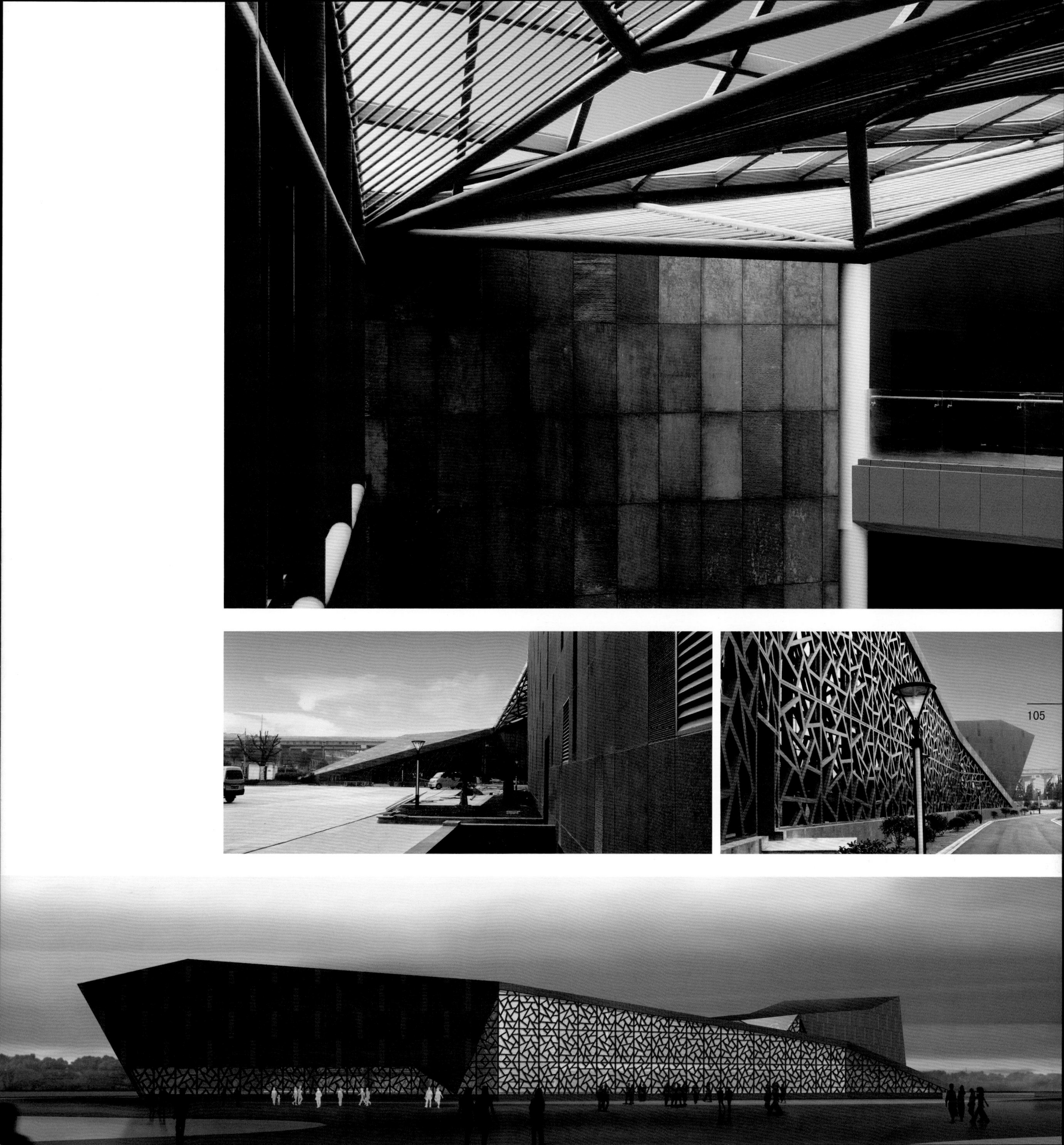

方案比较

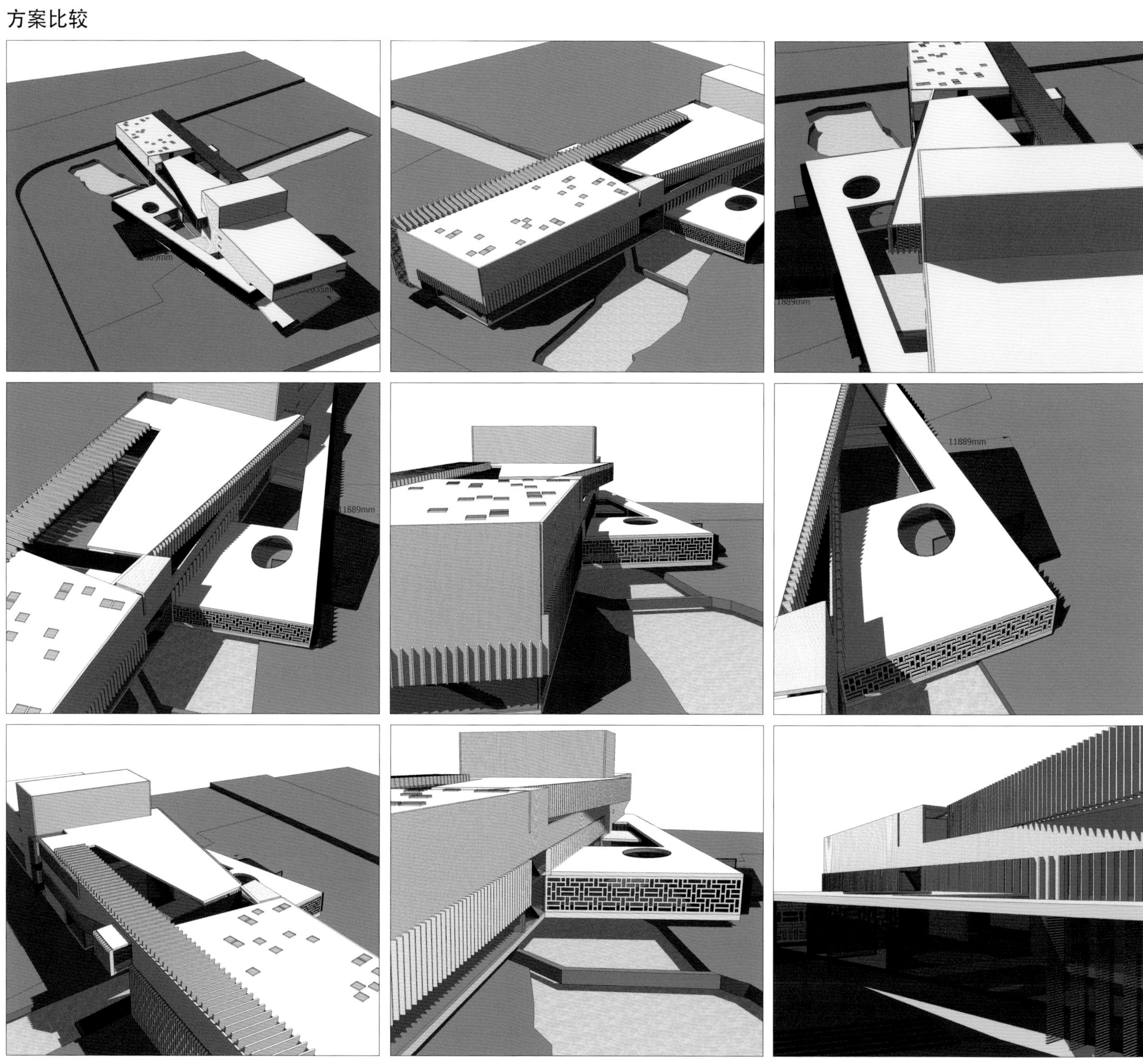

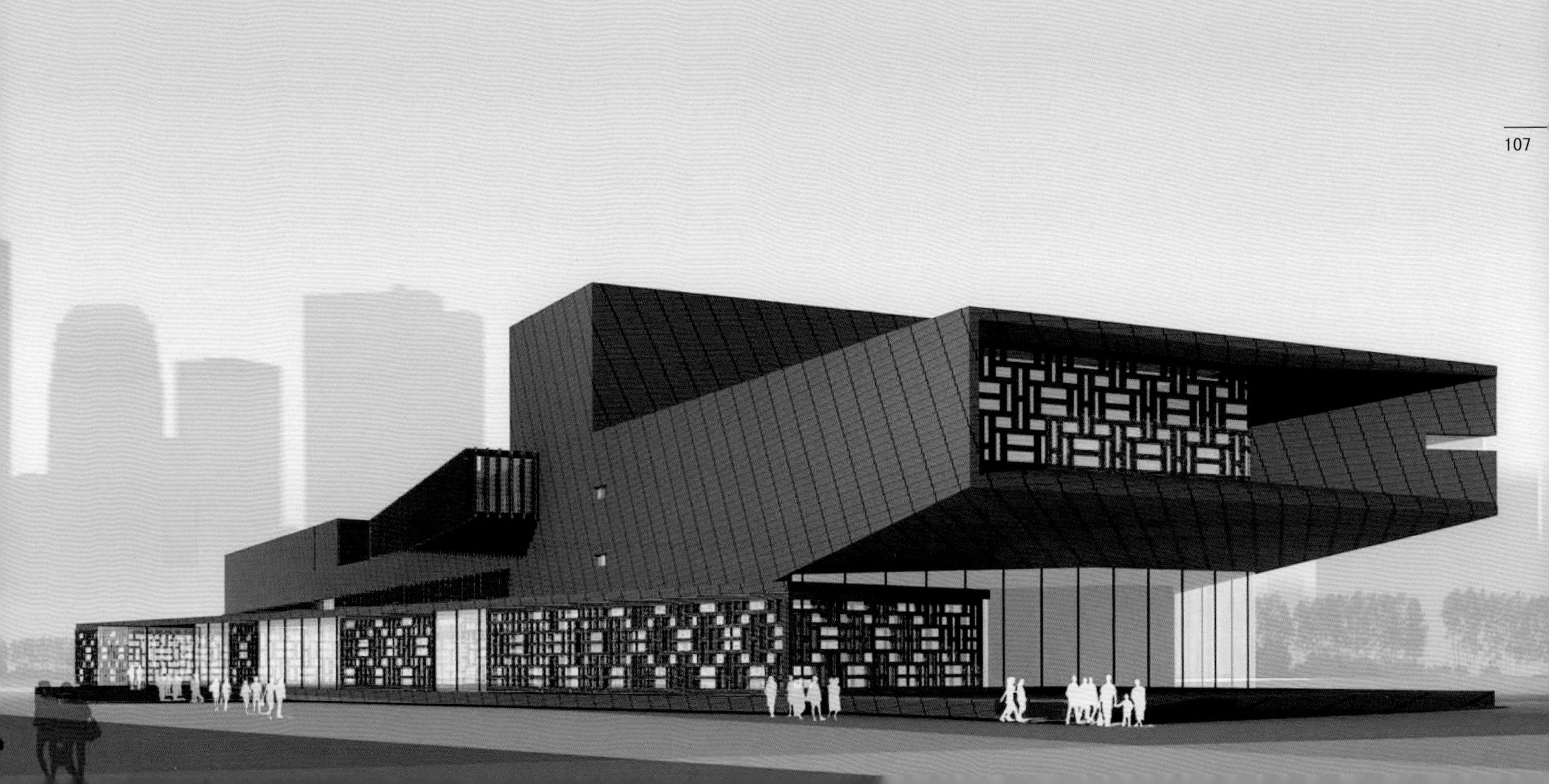

浙江省档案馆

Archives of Zhejiang Province

概 念 方 案　国内邀标 /2009
建筑师团队：高裕江、黄银金
合 作 单 位：杭州市建筑设计研究院

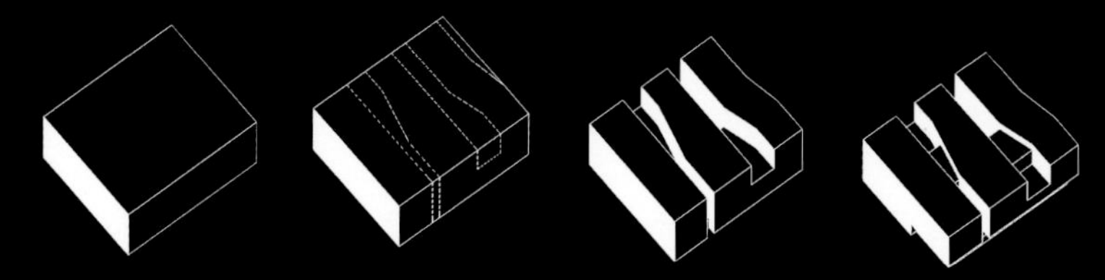

形体推敲

设计首先从形体把握及尺度控制上折射出现代建筑中所包含的地域建筑文化的意韵及文化基因，形似或神似“档案箱柜”的建筑形象具有象征的文脉因素，突显出生成建构的切合性。其次是建筑的虚实对比，纵横穿插以及相对搭接和承托，均具有艺术性和隐喻性。

Firstly, the design reflects the connotation and the cultural gene of regional architectural culture contained by modern architecture by grasp of form and control of scale, the architectural image of "file cabinet" has the symbolic context, highlighting the appropriateness of construction. Secondly, the comparison of virtual-real, vertical and horizontal thrust, the relative lap and support lap of architecture has a significant artistic and metaphorical nature.

总平面图 | Site Plan

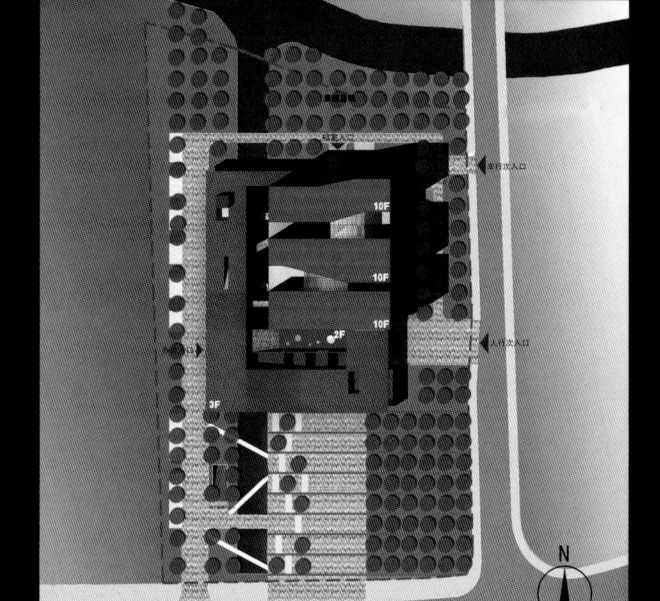

浙江省档案馆
Archives of Zhejiang Province

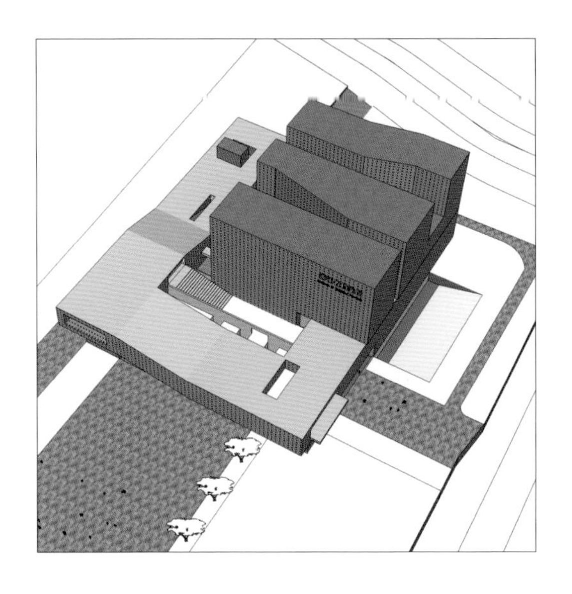
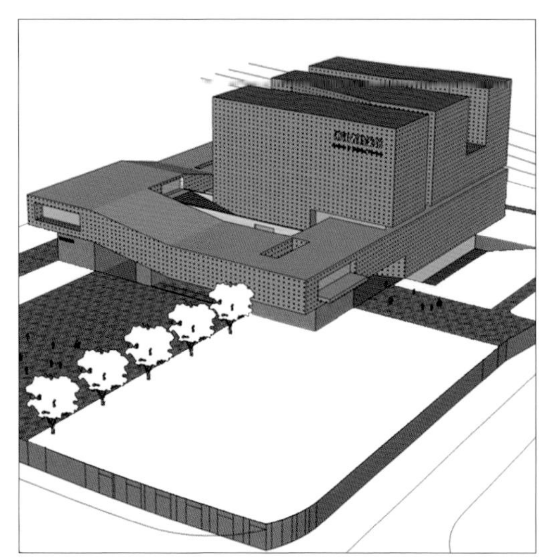
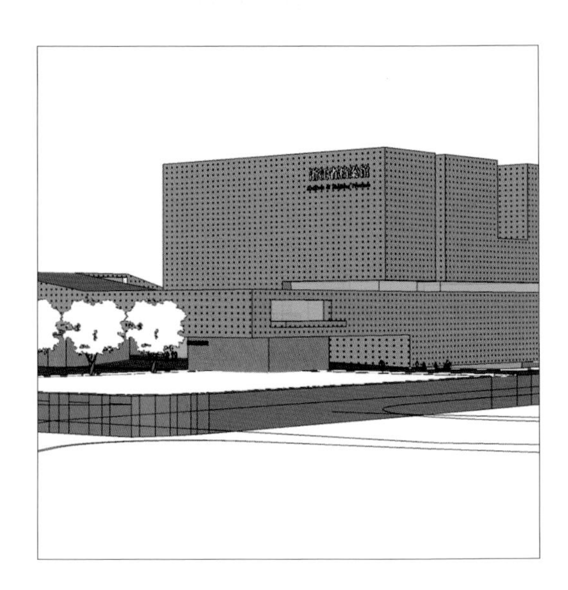
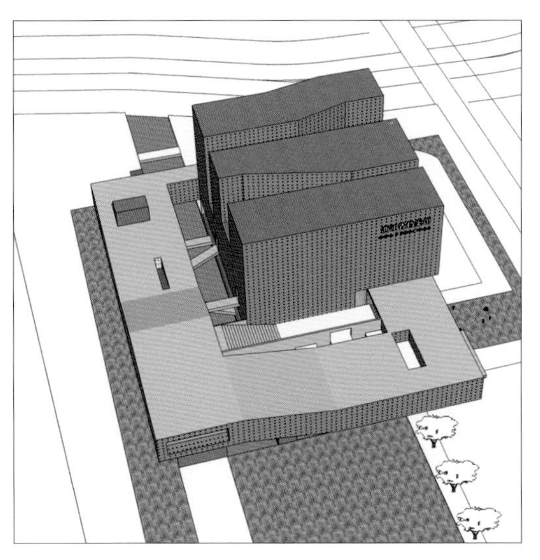
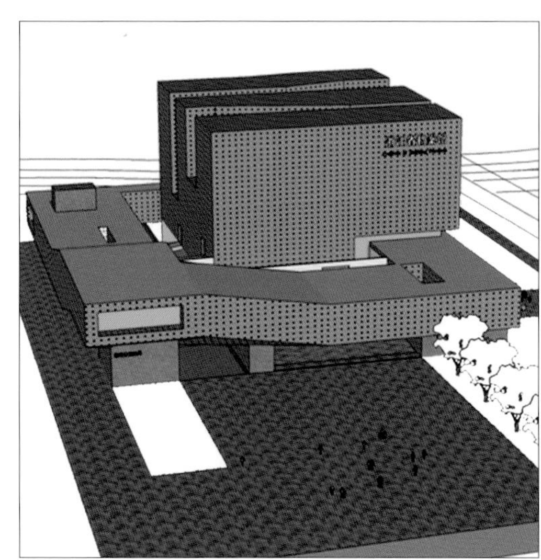
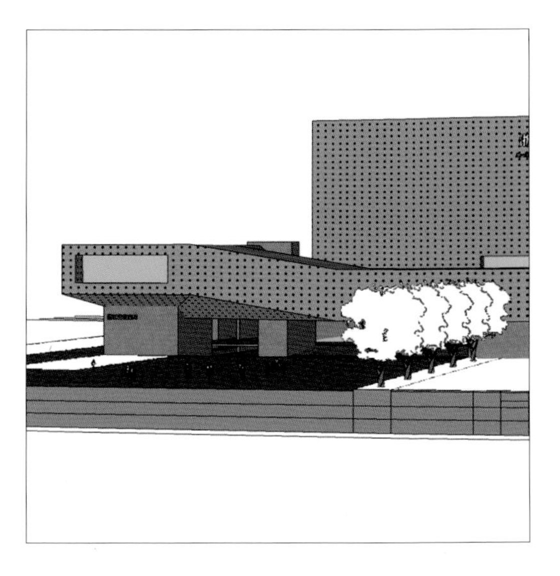

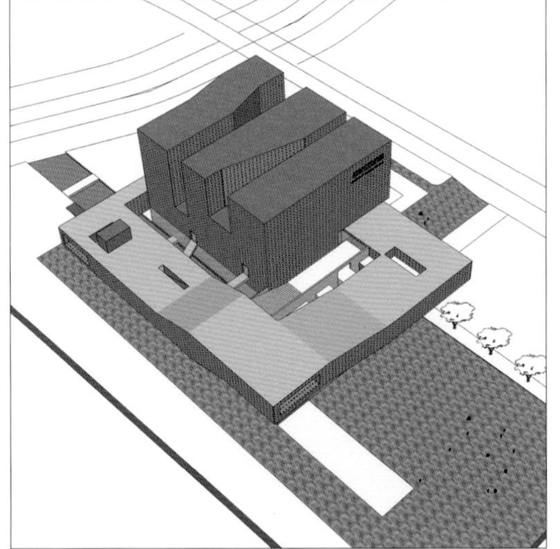
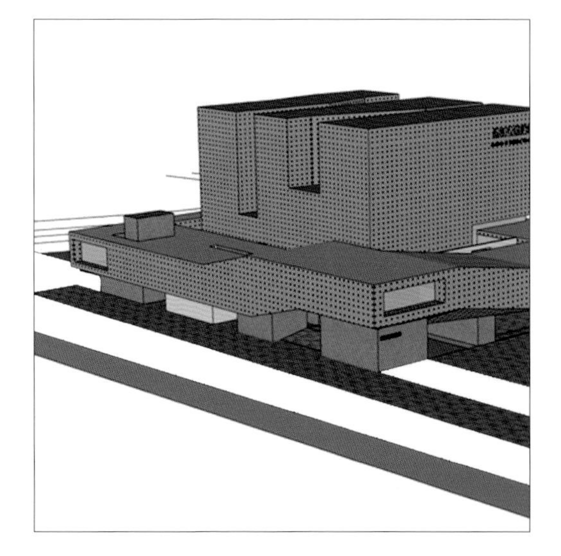
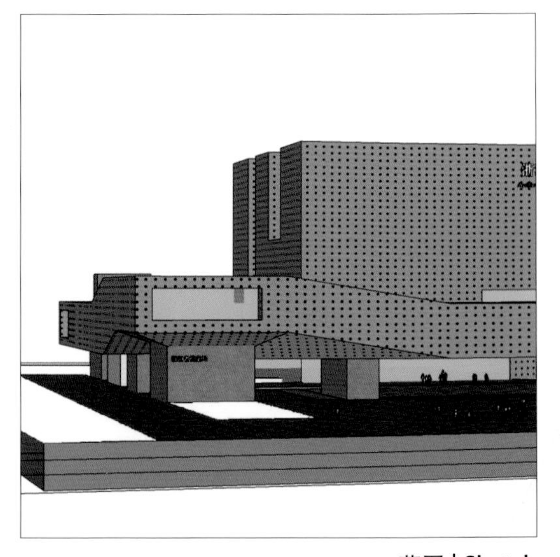

草图 | Sketch

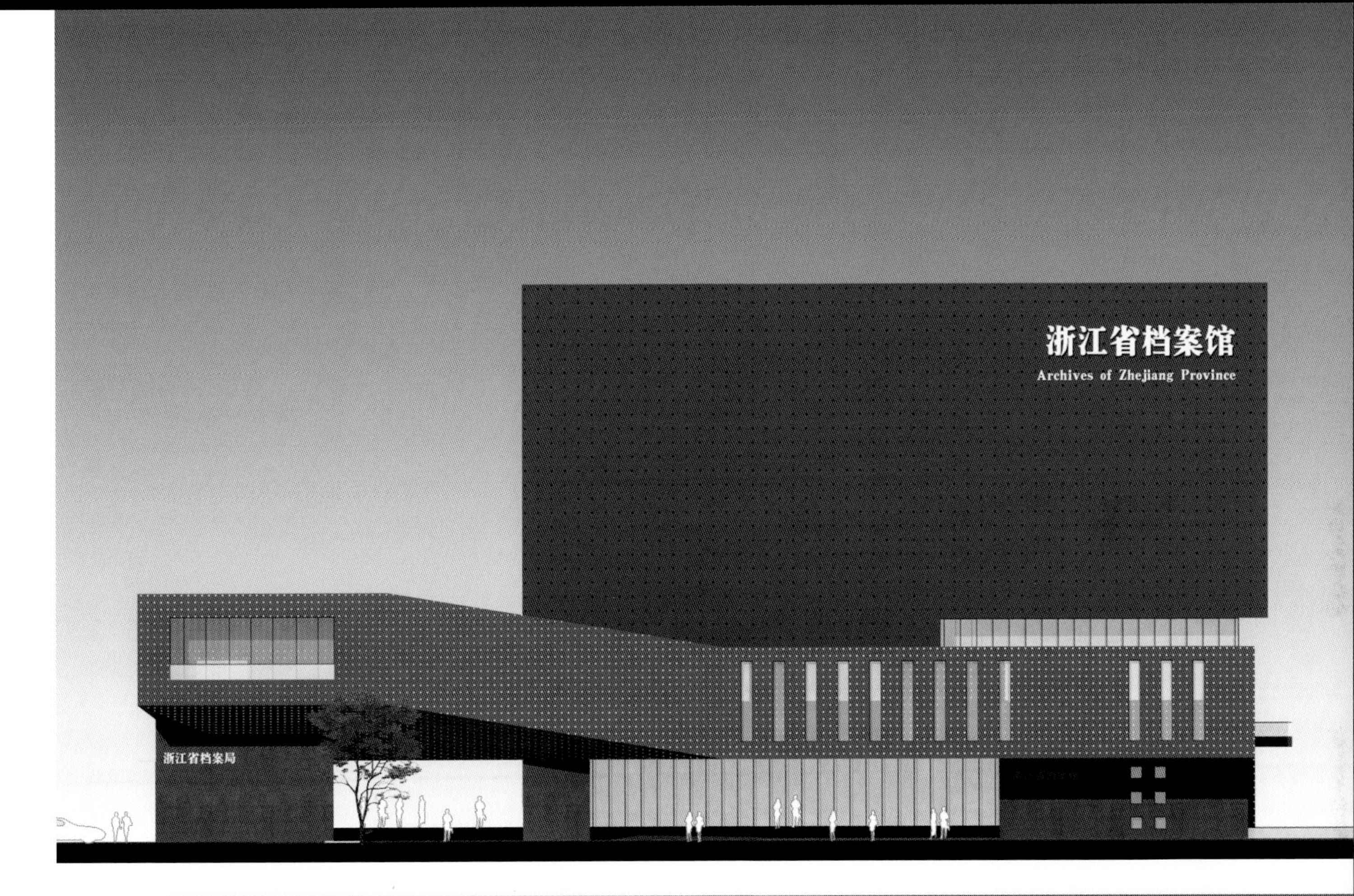

立面 | Elevation

浙江海洋学院

Zhejiang Ocean University

规 划 设 计　国内竞标 /2009
建筑师团队：高裕江、黄银金、亓茜、杨淼

设计充分体现三面临海一面靠山的优越自然基地条件，突出“海洋”要素；鉴于海塘围垦的低洼用地现状，以及环岛滨海景观大道的城市建设要求，使学校的教学生活既能达到安全、舒适、经济可行，并能借滨海特色风光之景，使城市滨海景观与高校校区的景观环境相整合；规划设计既运用了东方传统背山面水的理念，又能符合节能、节地、节水等绿色生态建筑的理念；建筑形态设计与长峙岛的欧洲传统小镇空间品质和意向的现代化海岛宜居之城相协调，形成现代大学校园开放、交流、融合的教学空间模式。

总平面图 | Site Plan

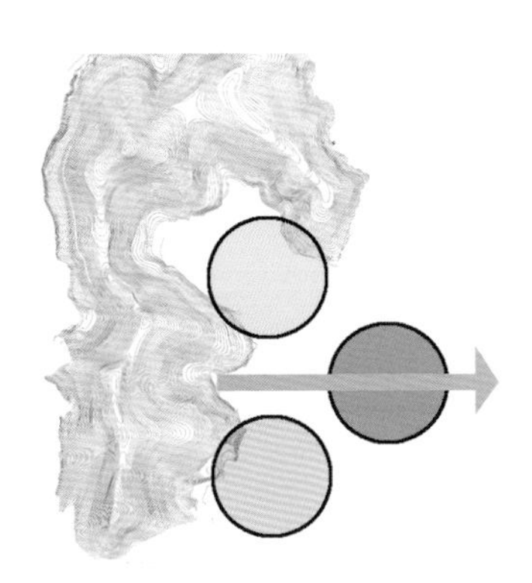

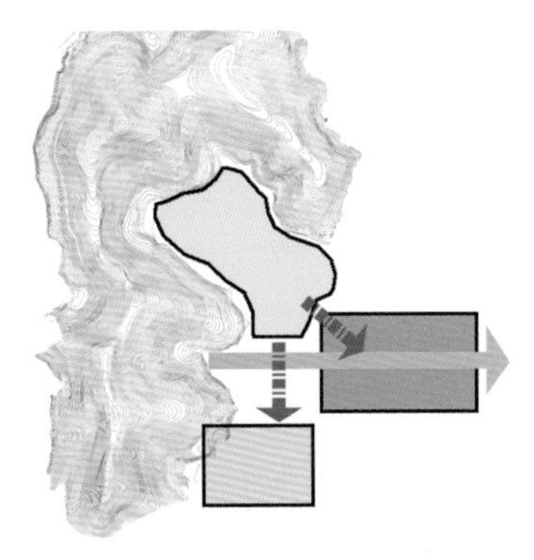

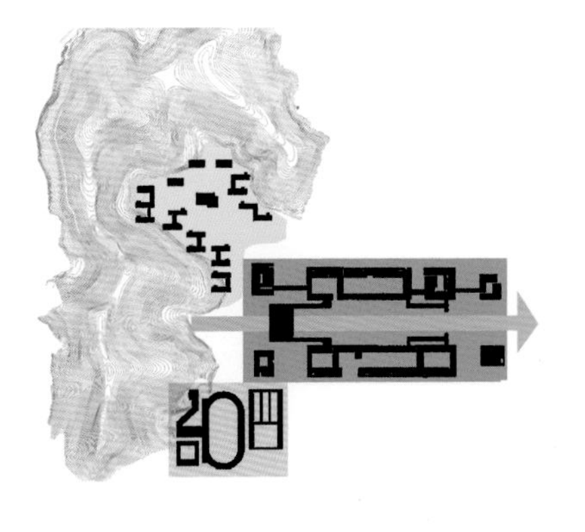

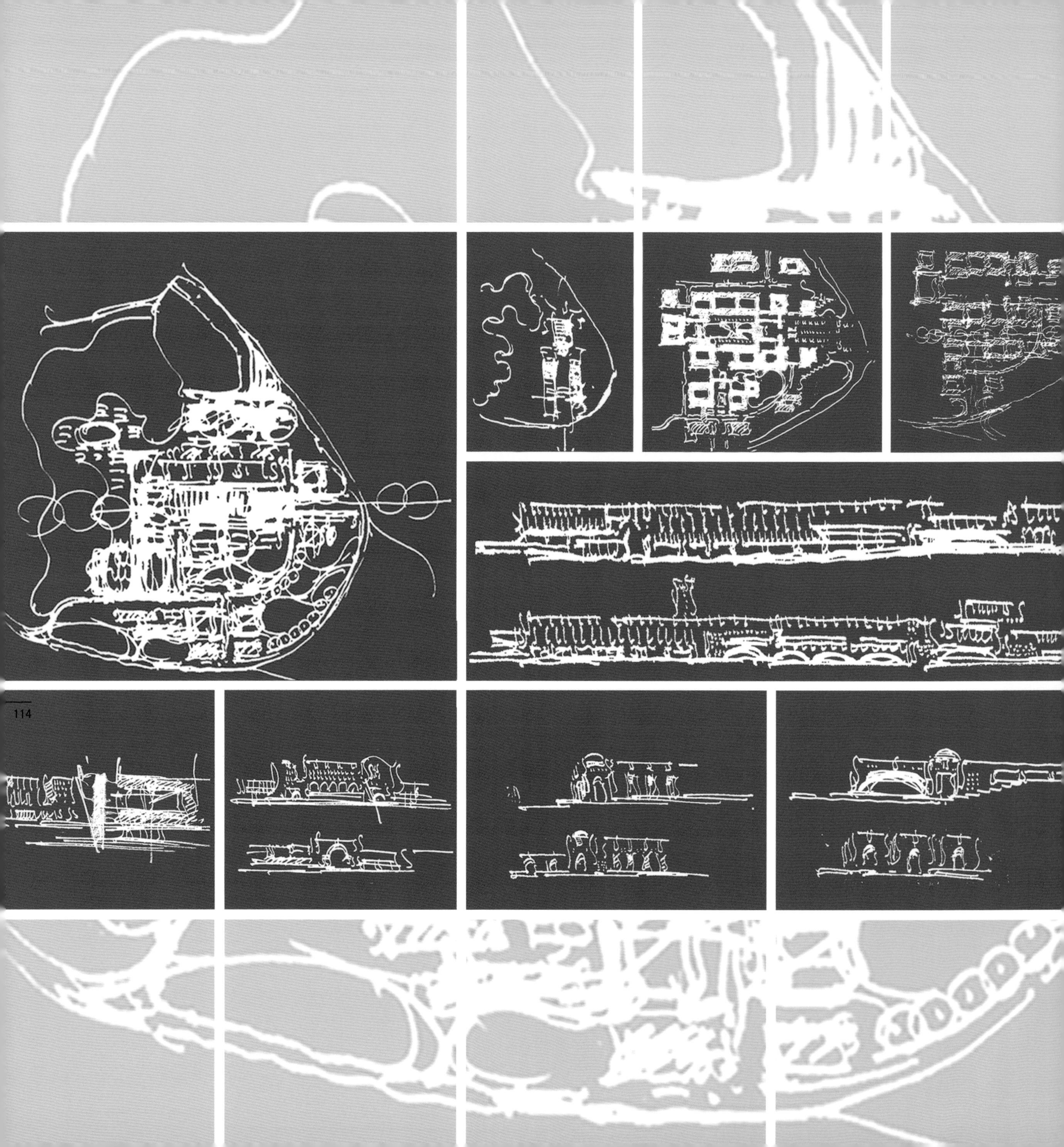

Design fully reflects the excellent natural conditions, which include the sea on three sides and mountain on one side. Highlighting the "marine" element. Design properly deals with the present situation of low lying lands and seaside landscape avenue around the island, to create a kind of safe, comfortable, economically feasible school life. Besides, integrating the landscape between the city seaside area and university campus is also important. Design makes full use of the plan concept of "backing on maintains and facing to sea" from oriental culture, and takes advantage of the good towards to meet the requirements of green ecological architecture including water,energy and lands resources saving. The other theme of the architecture spatial design is regional. Form design should coordinate to the European traditional town space and the city that is good for living on a modern island and can also meet the requirements of the teaching space in modern university which include opening, communication and inosculating.

20.0

60.0

剖面图 | Section Plan

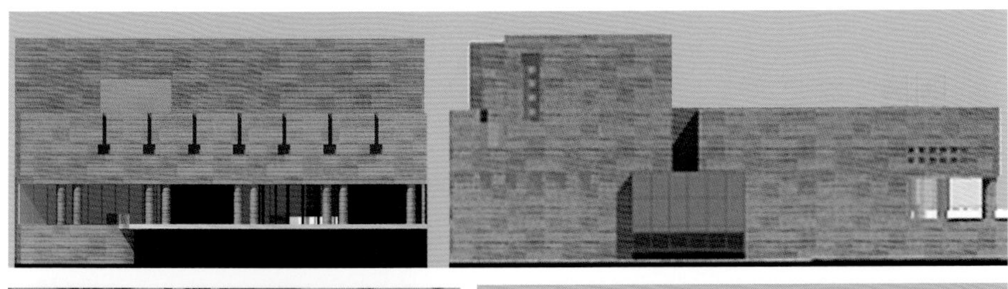

45.0

62.4

姜山镇文化体育中心及水街

Jiangshan Culture and Sports Center and Water Street

设 计 方 案　2009
建筑师团队：高裕江、黄银金、严聪

设计运用几何、理性的手法，使建筑内在功能和形象特色有机融汇，同时承传江南民居的形态模式，架构出一组既经典雅致，又显清新简约，并融传统性、文化性、艺术性、时代性于一体的现代标志性建筑。

This design adopts a geometrical and rational way, making an organic combination between the building's intrinsic function and extrinsic image characteristics. At the same time, inheriting the morphological pattern of Jiangnan(regions south of the Yangtze River in China) folk house, the designer constructs a set of modern landmark buildings with the characteristics of classical, elegant, pure and terse, which integrates tradition, culture, artistry and modernity.

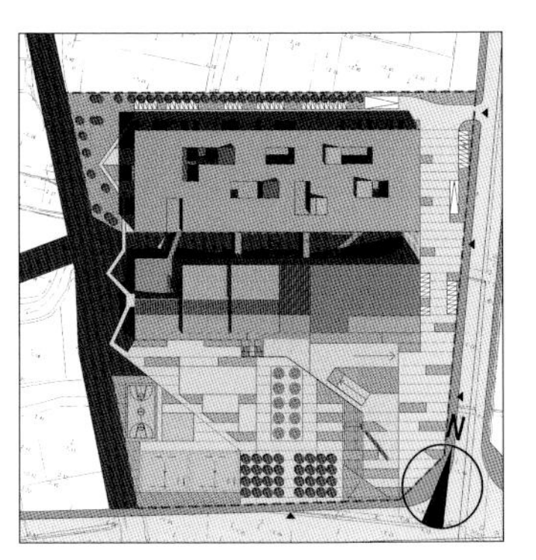

总平面图 | Site Plan

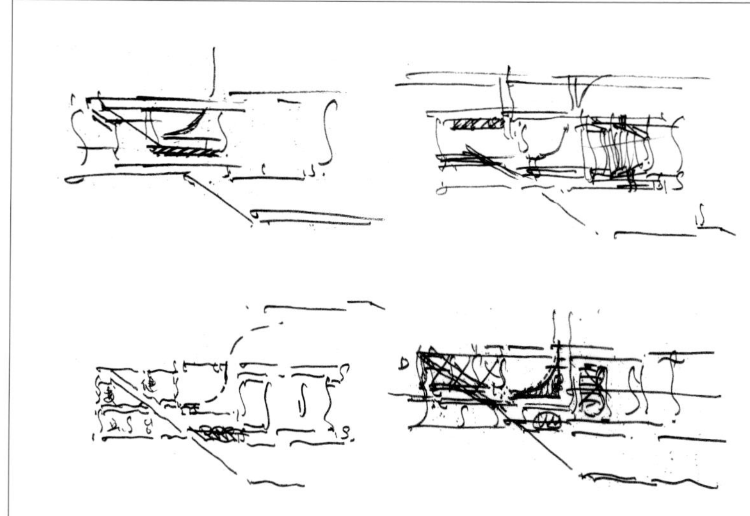

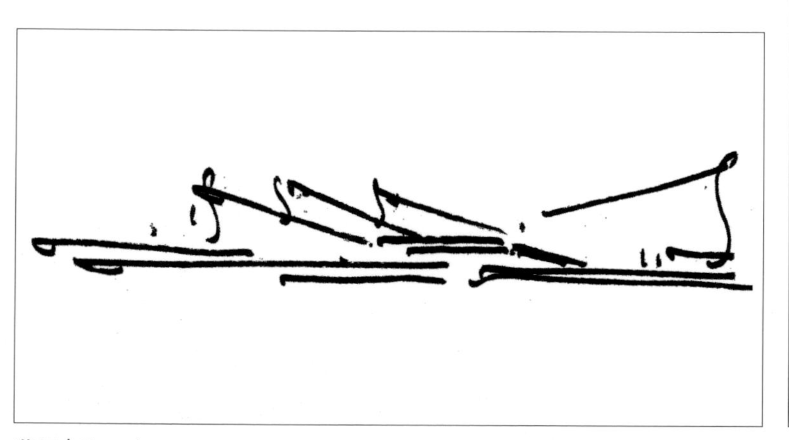

草图 | Sketch

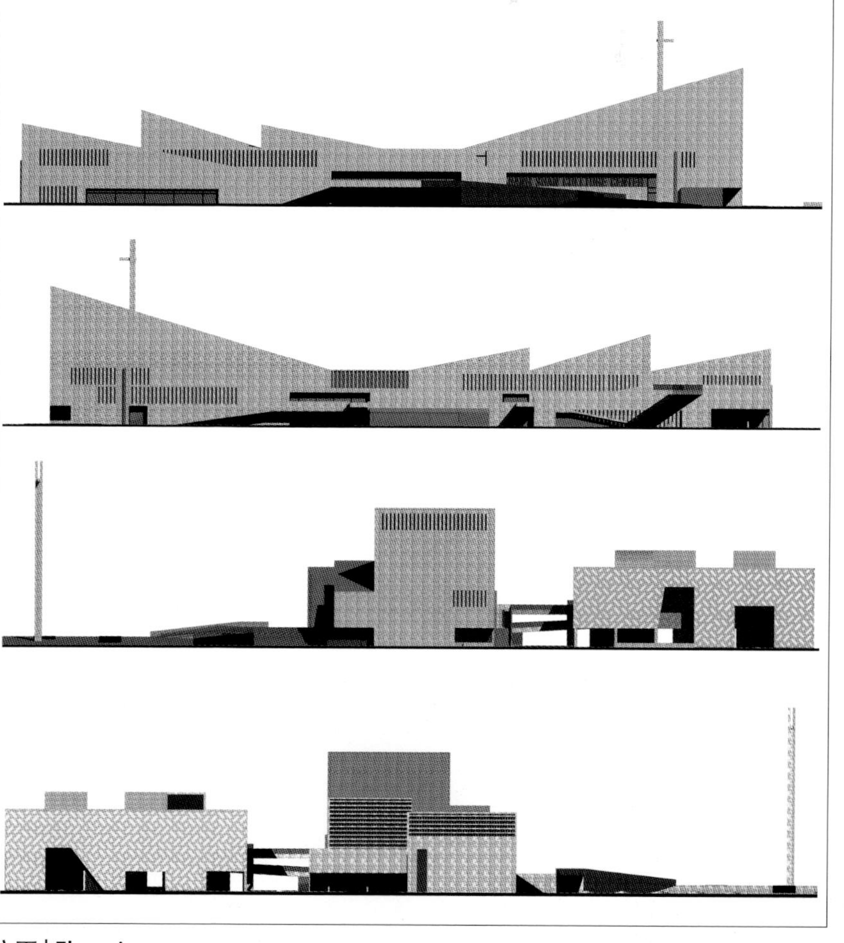

立面 | Elevation

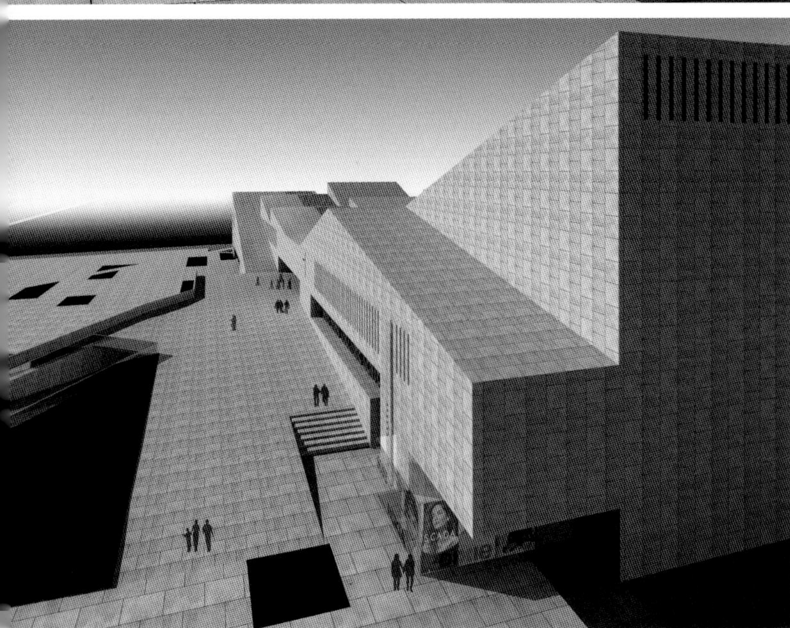

天津图书馆

Tianjin Library

设 计 方 案　国际邀标 /2009
建筑师团队：高裕江、杨淼、顾月明、周敏、陆卫兵、刘娜、陆信东

图书馆形象构思来源于“新旧知识的融汇与叠加，过去与未来的融合和相通，古代与今世交融和对话”，也象征着“古典书籍”与“当今电脑”的错时重叠。

以深褐色穿孔板金属百叶遮阳构件面与洋红木色的金属板窗洞缺口组成的半开启“古书”置于基地东侧，形成图书馆的“基座”，而银灰色的铝板“书简体”饰面以倒置的形式架于基地的西侧，并与“古书”相扣，形似笔记本电脑，其表面的遮阳体以古书简的形式，不仅传递着古幽的文化气息，而且其屋顶太阳能光伏电子板与建筑顶面一体化的处理，显示出科学理性的精神。

The library design concept is derived from "the integration of the new and the old knowledge, the fusion of the past and future, and the combination of the ancient and the present"; Furthermore, it indicates the overlapping of the "ancient book" and "present computer".

The dark-brown perforated metal adumbral louver component and the carmine mental hole are combined to form the half-open "ancient book" on the east of the site, so as to build up the "base" of the library. In addition, the laptop-shaped silver-gray aluminous object with "simplified character" is inversely located on the west of the premise and connected with the "ancient book". The adumbral item adopting the form of ancient correspondence presents the national culture; the combination of the solar energy panel and the roof shows the spirit of scientific rationality.

概念草图 | Concept Sketch

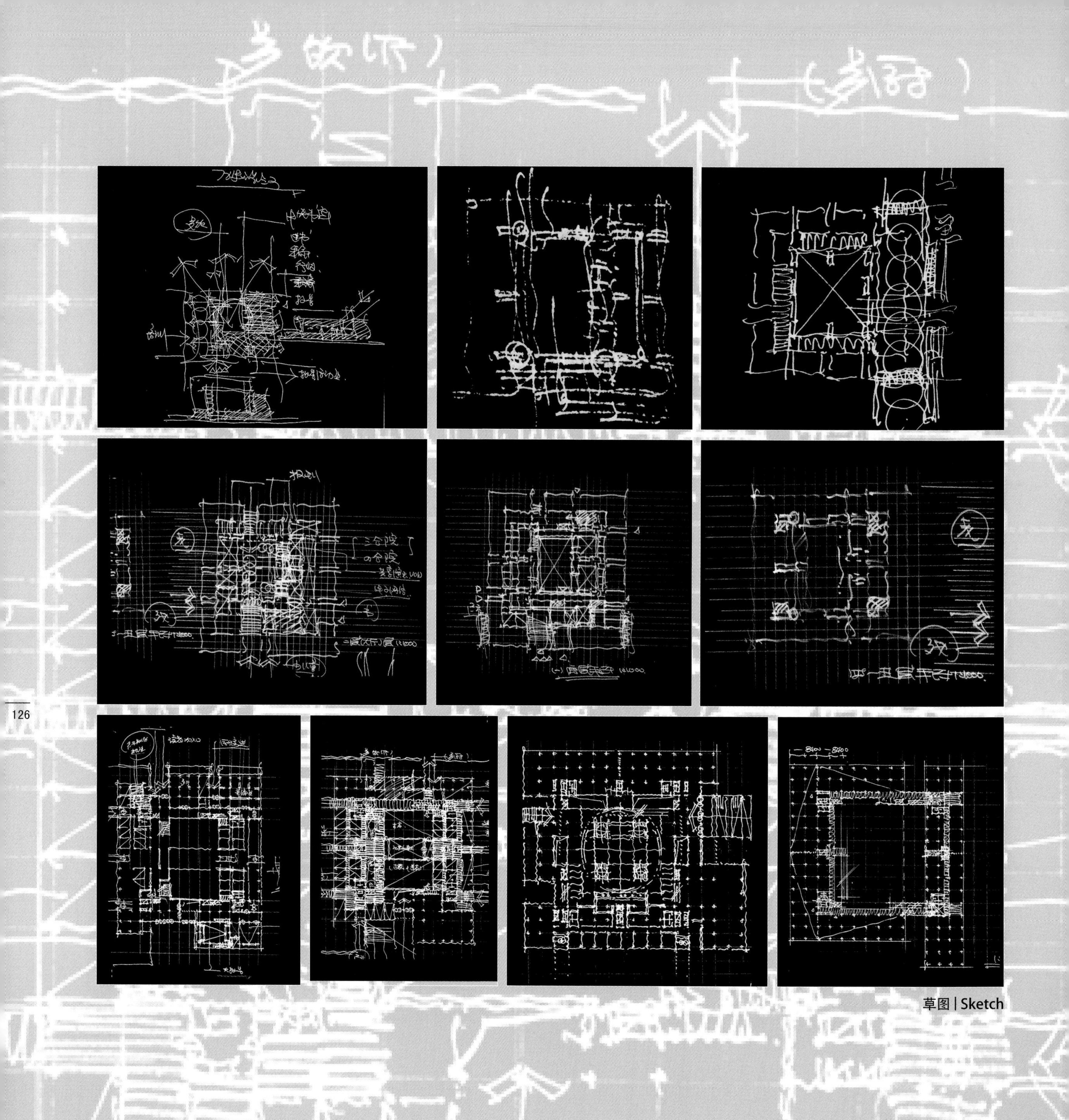

草图 | Sketch

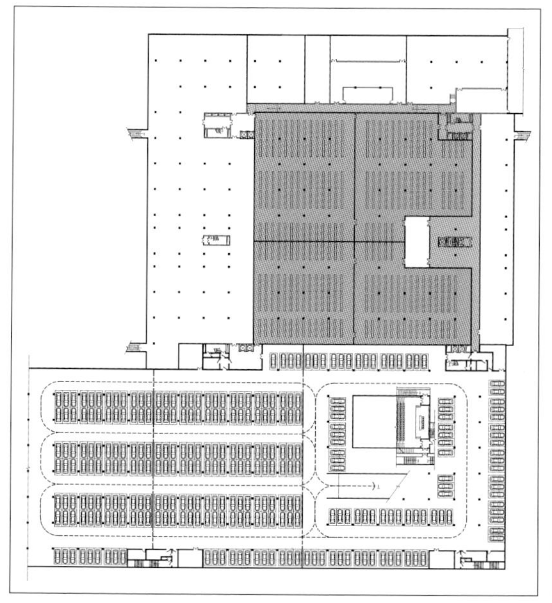

地下一层平面 | Basement Floor 1 Plan

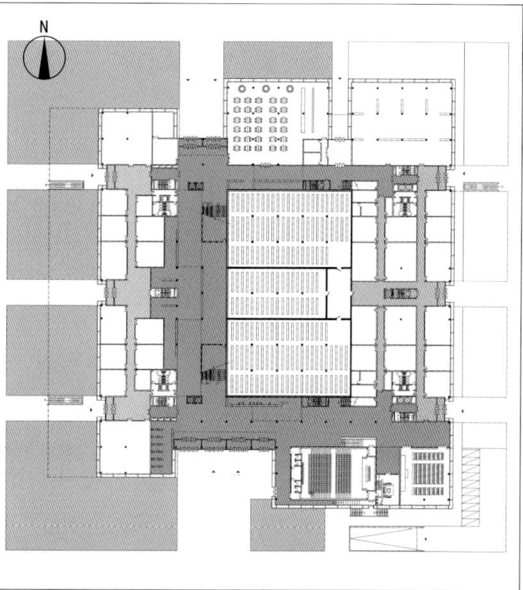

一层平面 | Ground Floor Plan

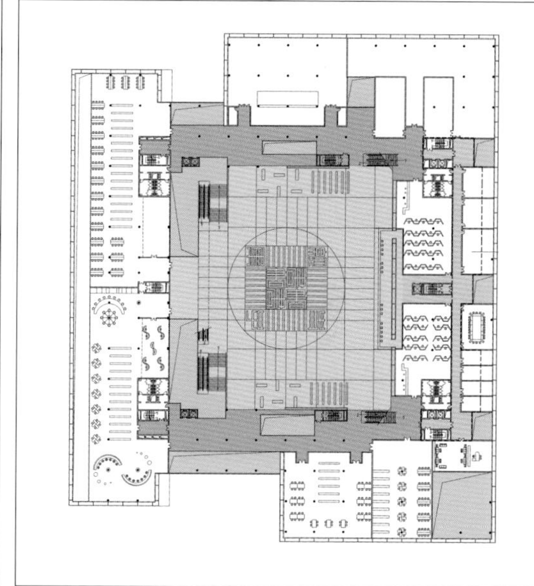

二层平面 | Second Floor Plan

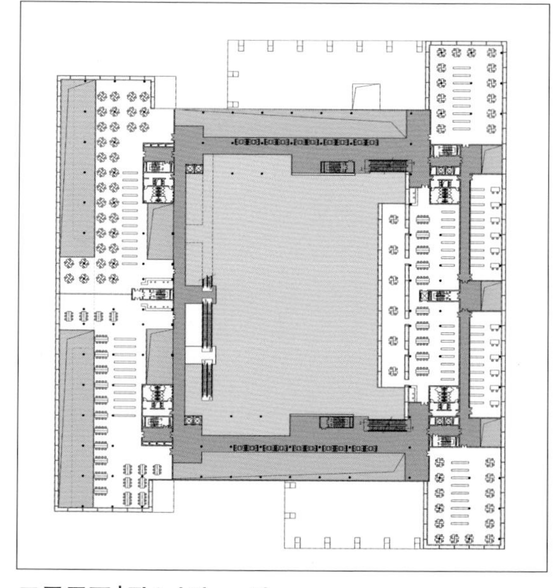

三层平面 | Third Floor Plan

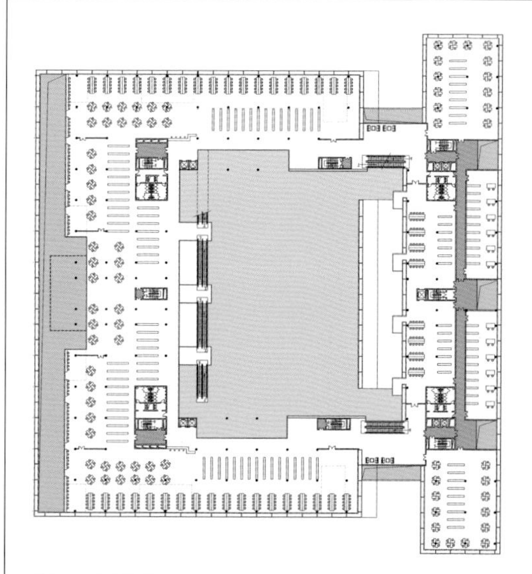

四层平面 | Fourth Floor Plan

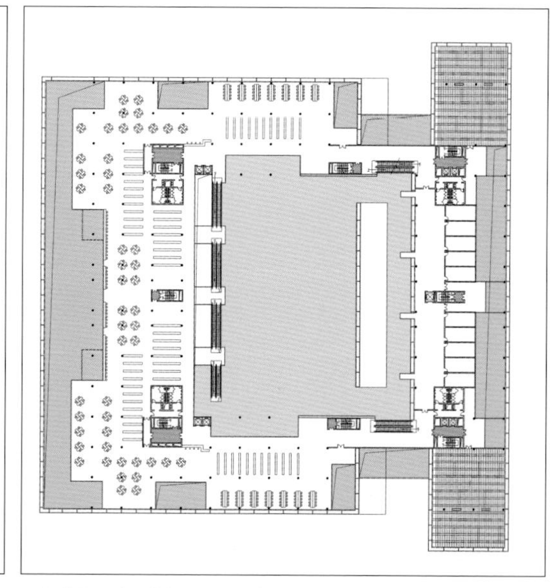

五层平面 | Fifth Floor Plan

功能分析

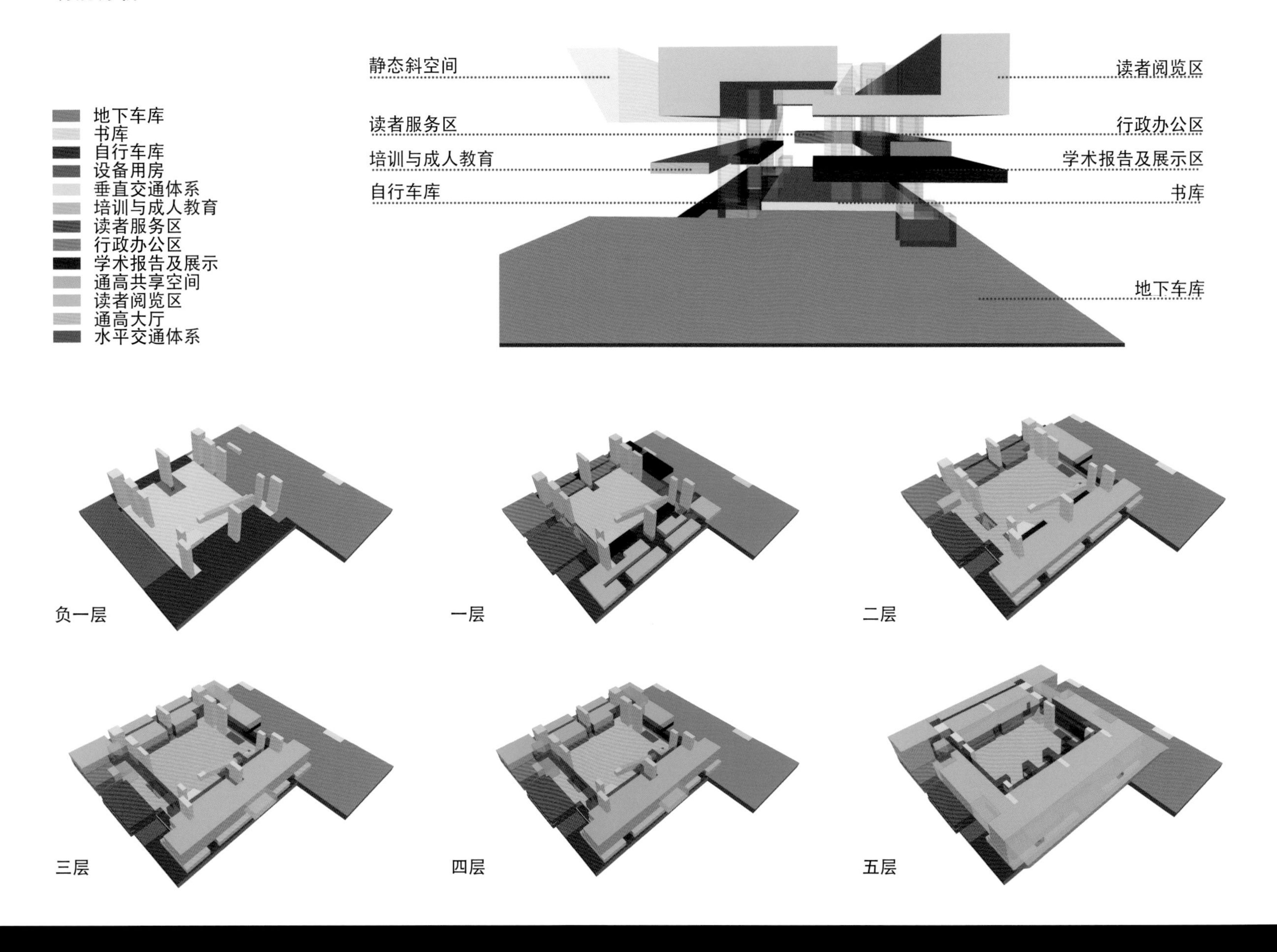

地下车库
书库
自行车库
设备用房
垂直交通体系
培训与成人教育
读者服务区
行政办公区
学术报告及展示
通高共享空间
读者阅览区
通高大厅
水平交通体系
静态斜空间
读者服务区
培训与成人教育
自行车库
读者阅览区
行政办公区
学术报告及展示区
书库
地下车库
负一层
一层
二层
三层
四层
五层

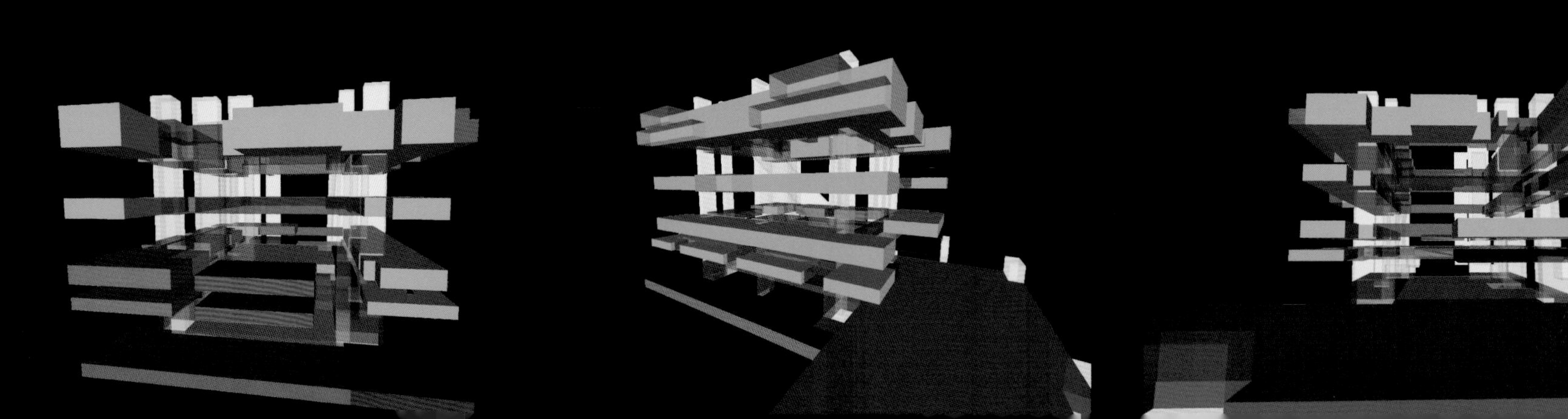

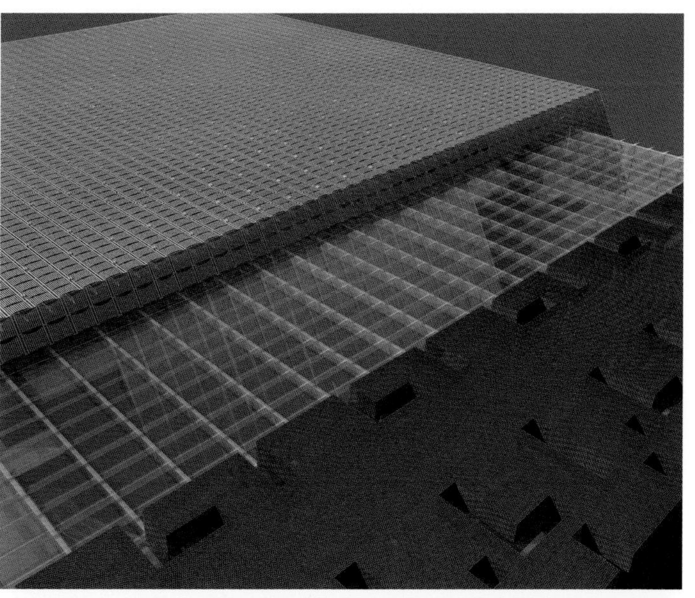

细部 Detail

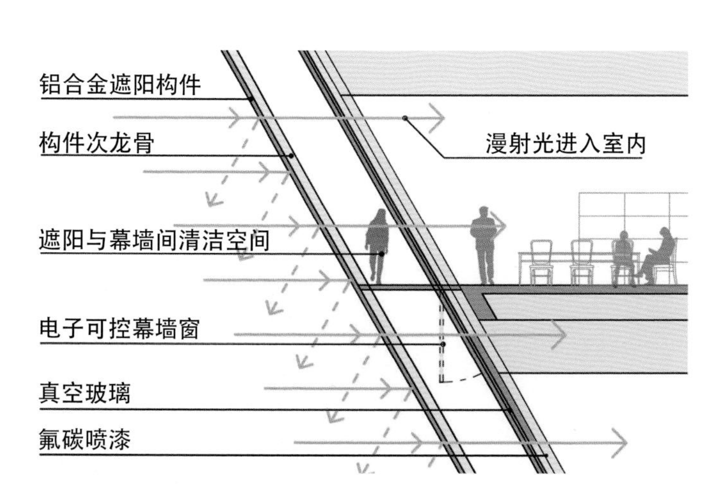
铝合金遮阳构件
构件次龙骨
漫射光进入室内
遮阳与幕墙间清洁空间
电子可控幕墙窗
真空玻璃
氟碳喷漆

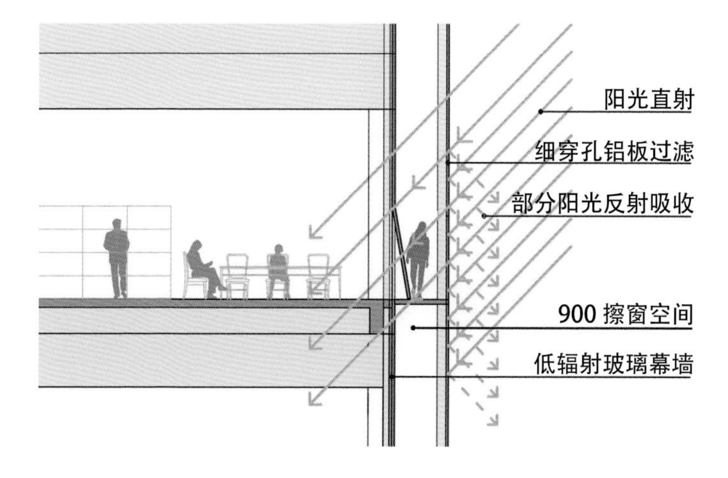
阳光直射
细穿孔铝板过滤
部分阳光反射吸收
900 擦窗空间
低辐射玻璃幕墙

中国银行南京运营中心

Bank of China Operations Center in Nanjing

设 计 方 案　国际竞标（第一名）/2009
建筑师团队：高裕江、黄银金、毛志远、贾茜、许小笛

设计采用“轴廊、桥廊”的设计手法，将各功能区域合理有机地整构起来，从而形成科学合理的功能、交通形态模式。运用传统院落空间形制，结合运行办公、培训教学对建筑自然采光、通风、节能等空间生态性、景观性的内在需求，建构出空间大小不一、形态各异、封闭与开敞相融的各类院型空间，传承优秀的传统建筑文化，营建舒适、优雅的工作、教学环境。

Designed with the "Axis Gallery, Bridge Gallery" approach, each functional area are integrated reasonably and organically to form a scientific and rational function, traffic patterns. By making use of traditional courtyard space system, in conjunction with internal demand of functional area for building natural lighting, ventilation, energy saving and so on, the campus constructs various types of courtyard spaces of different sizes and shapes in combination of being closed and opened, so as to inherite excellent traditional architectural culture and create a comfortable, and elegant environment for work, teaching and learning.

草图 | Sketch

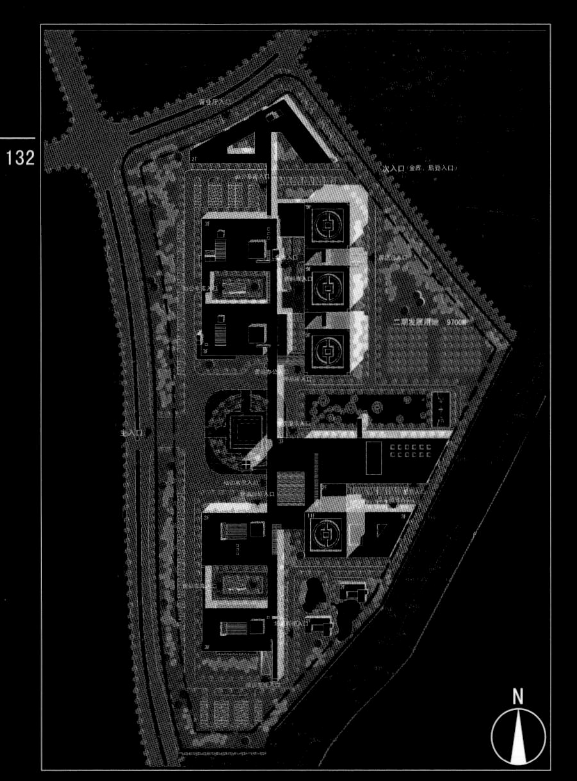

总平面图 | Site Plan

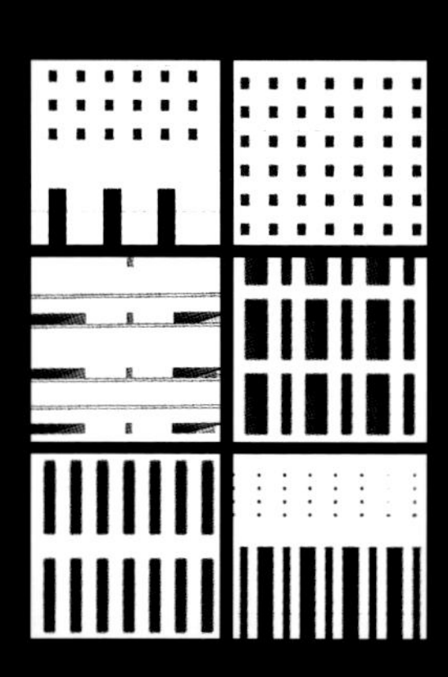

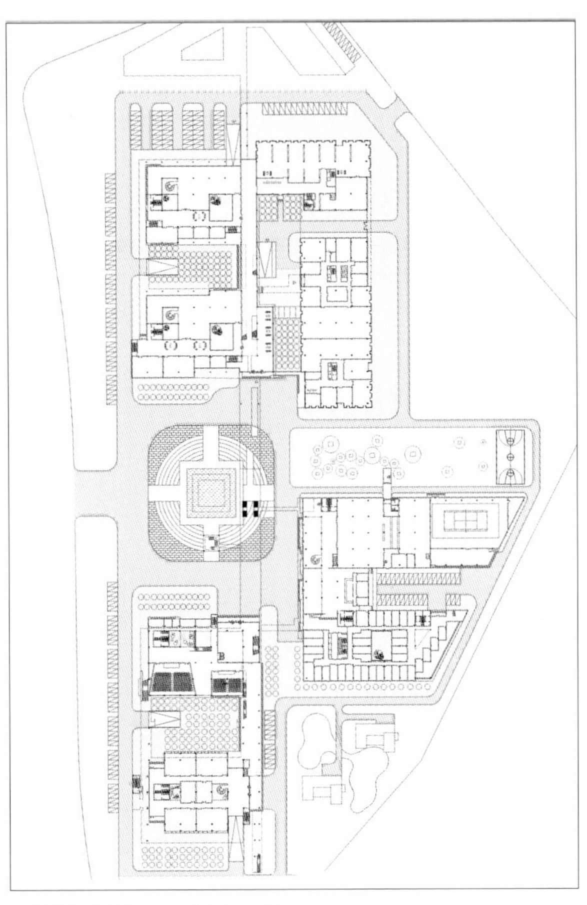

一层平面 | Ground Floor Plan

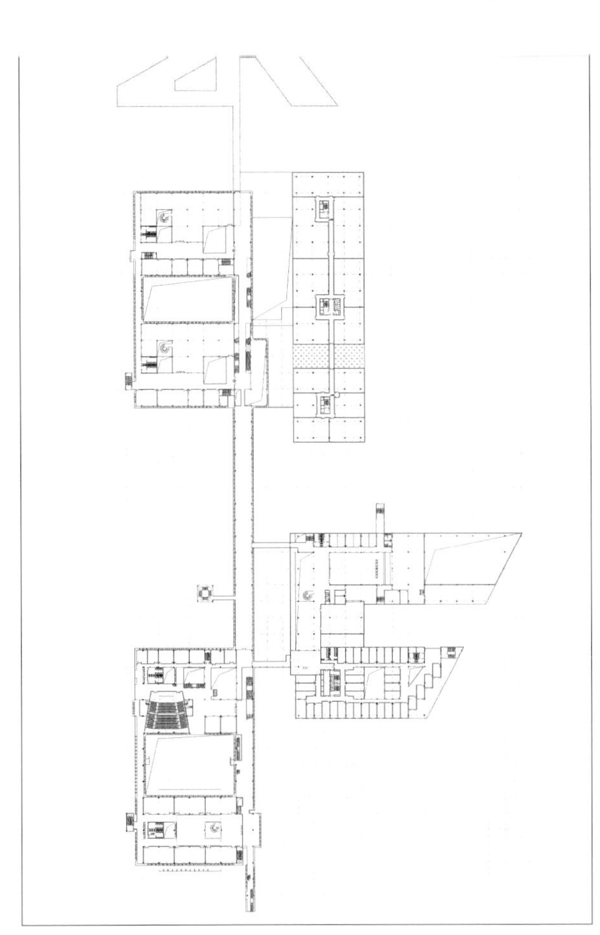

二层平面 | Second Floor Plan

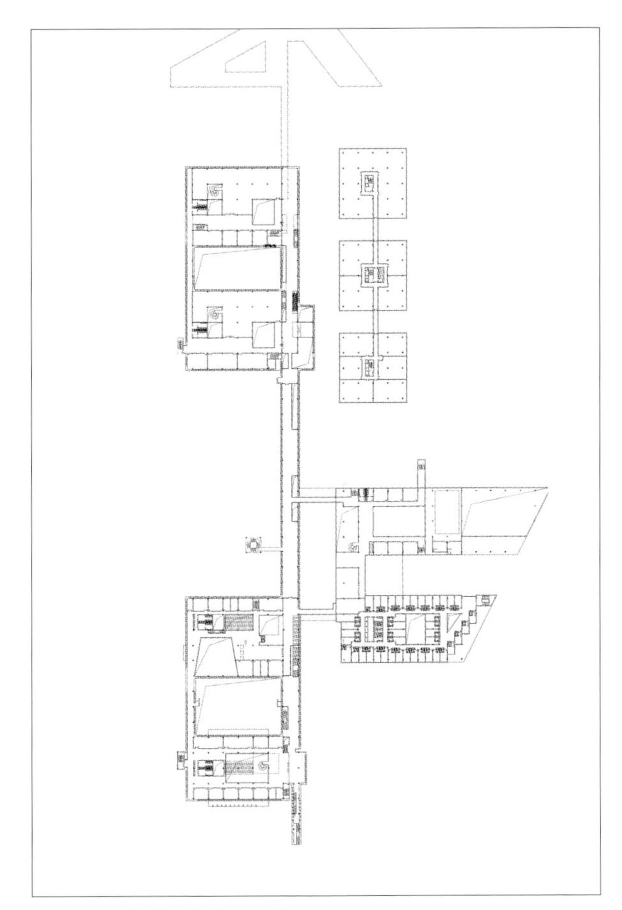
三层平面 | Third Floor Plan

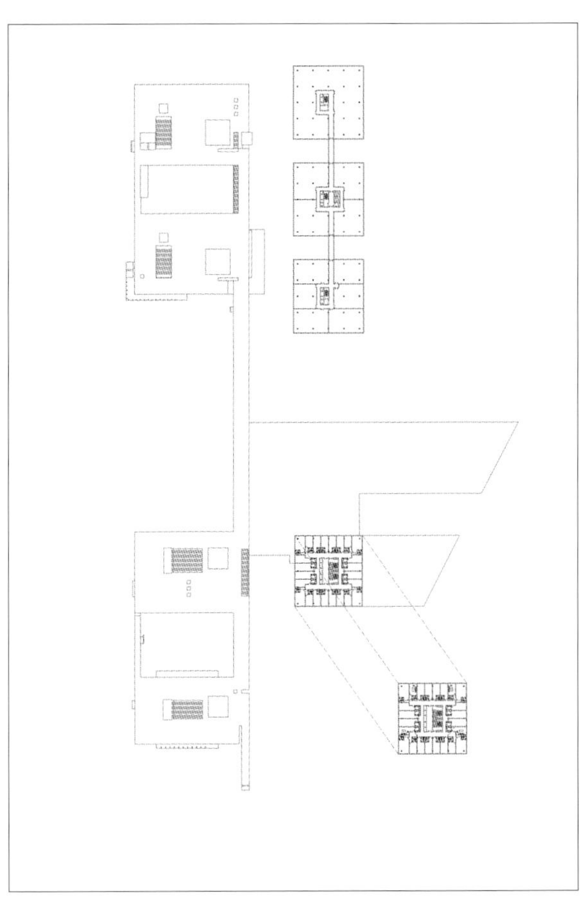
四层平面 | Fourth Floor Plan

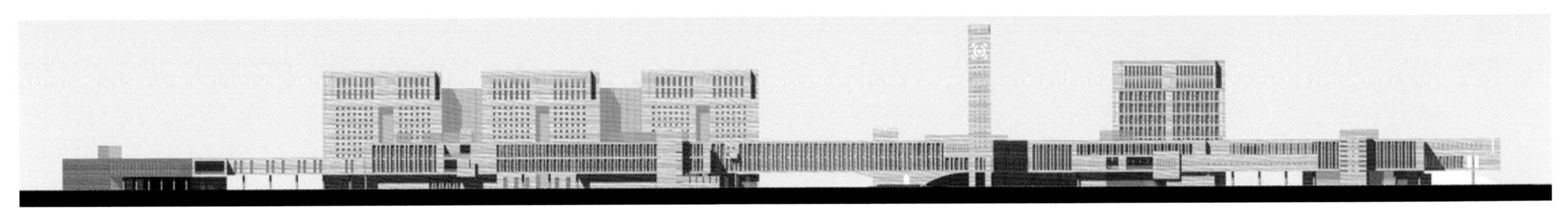
西立面 | West Elevation

东立面 | East Elevation

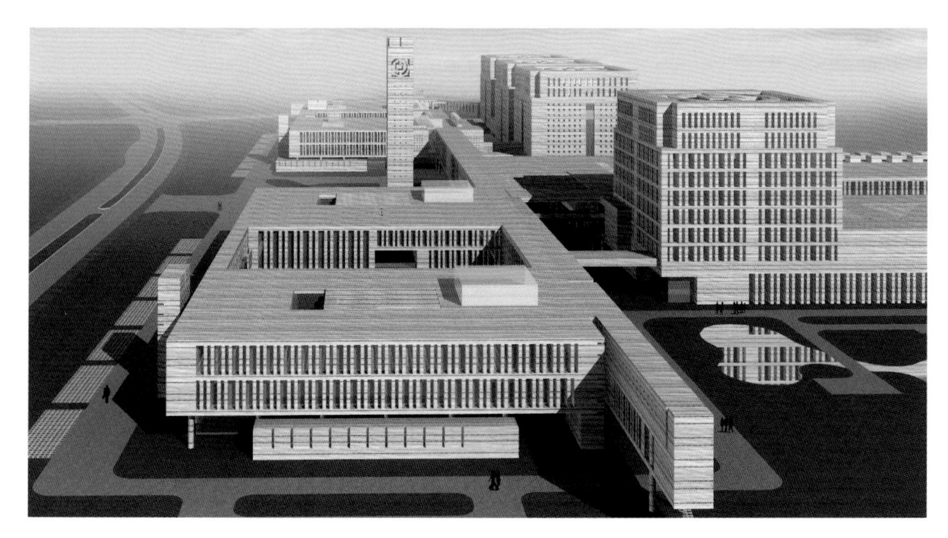

宁波鄞州中学体育中心

Sports Center of Yinzhou Middle School, Ningbo

实　施　工　程　国际竞标（第一名）/2010—2015
建 筑 师 团 队：高裕江、王明霞、蒋君标、沈晓鸣、史国雷、郭宁、赵鑫、饶峥、马云飞、刑明泉等
施工图合作单位：宁波市城建设计研究院

宁波鄞州中学的校园总体布局以教学楼群的整体轴网构架与横穿校园基地的原有河道整合互动为基础，形成了融建筑群、院落空间及滨水河道景观于一体的建筑形态格局。校体育中心正是在这一格局中的一个空间节点。在这种现实条件下，必须贯穿的河道将体育中心用地划分成两块不等的用地，反过来，也就是体育中心成为了跨越河道的建筑——成为非常独特的环境形态格局。因此，建筑形式自然同“拱桥”有关联，也具有了某种富有动感的建筑形象成为一种较自然的选择。这种楔入建筑特有场地所形成的建筑形态特征，展示的或许正是建构的一种逻辑性和科学性。

Ningbo Yinzhou Middle School, the basis for the overall campus layout is the integration of integral axial network of teaching building complex and the original watercourse crossing, thus producing the building format layout integrating building complex, courtyard space and riverside watercourse landscape. The Sports Center is a space node inside this pattern. Under these practical conditions, the watercourse divides building format naturally related with "arch bridge", with dynamic building image. The architectural morphological characteristics fit for the specific building site which display some sort of logicality and scientificity.

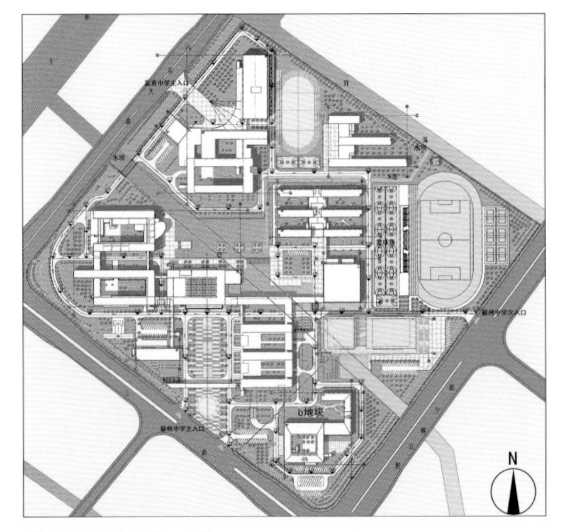

总平面图 | Site Plan

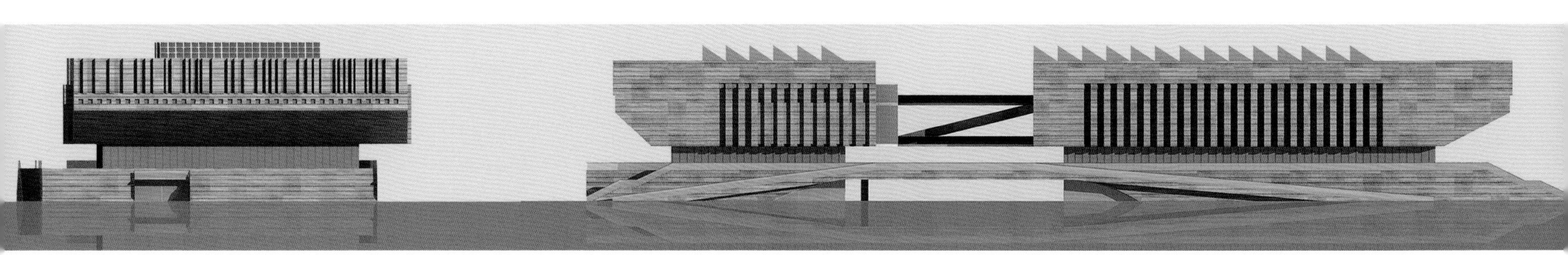

东立面 | East Elevation

南立面 | South Elevation

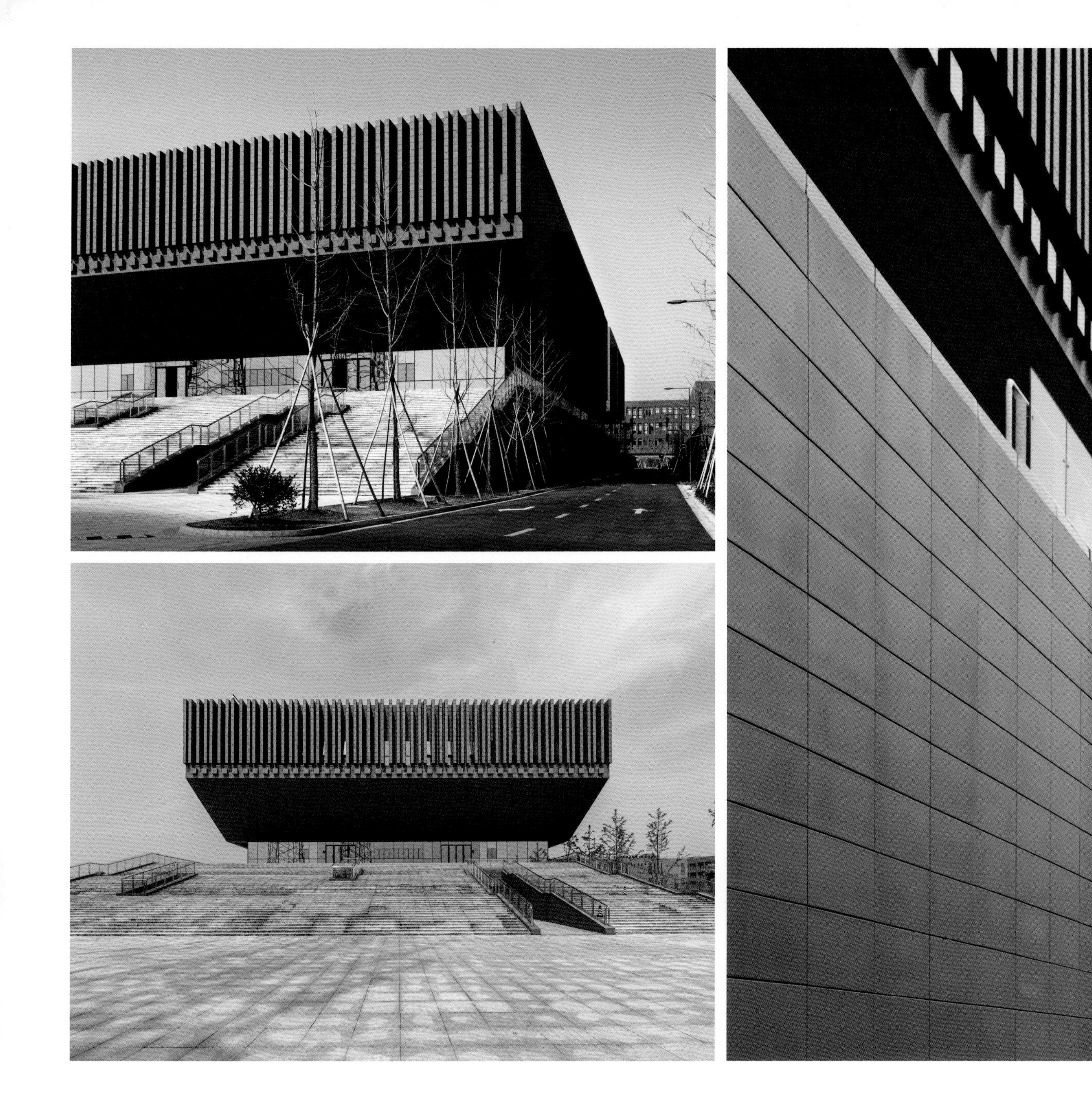

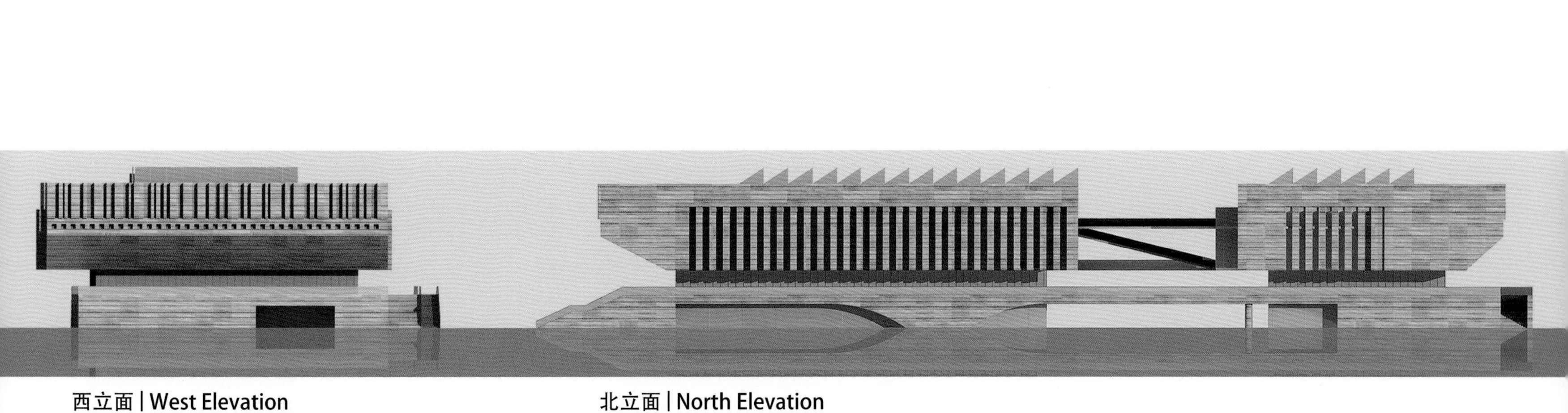

西立面 | West Elevation

北立面 | North Elevation

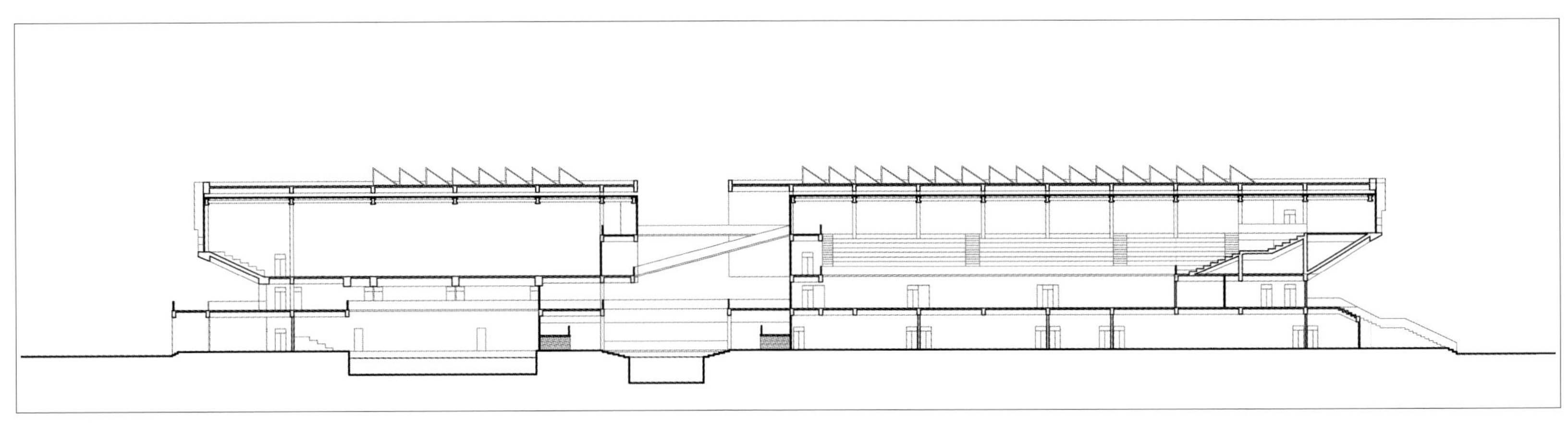

剖面图 | Section Elevation

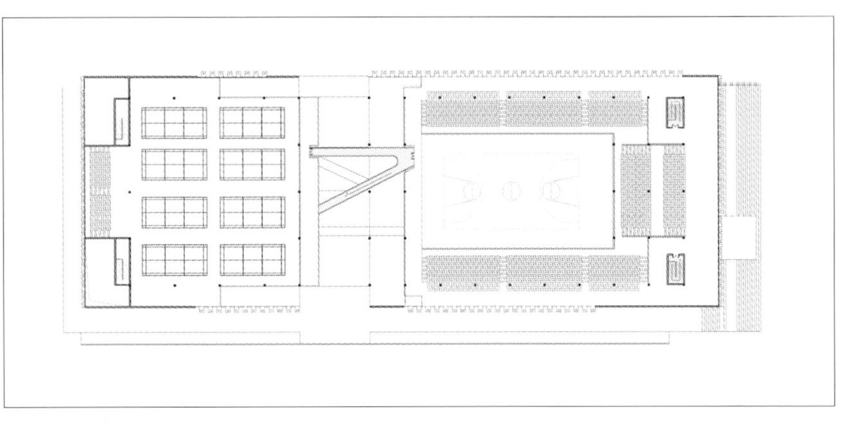

四层平面 | Fourth Floor Plan

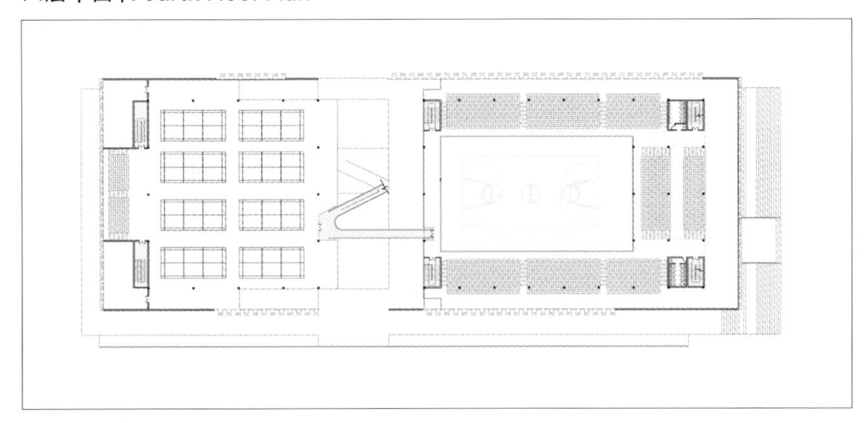

三层平面 | Third Floor Plan

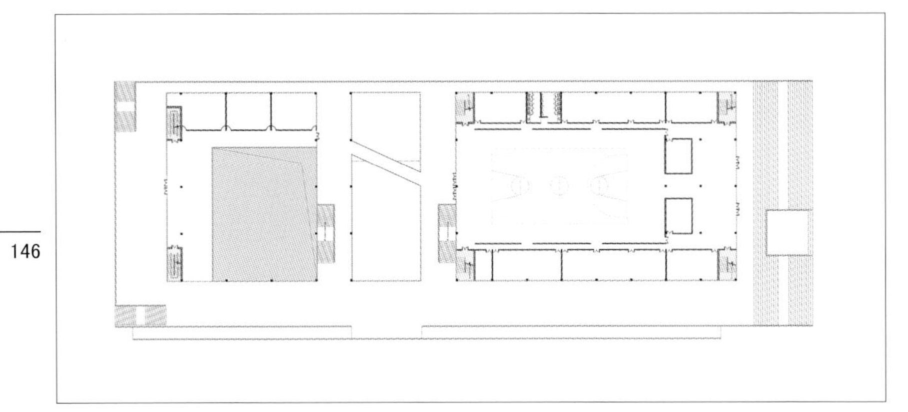

二层平面 | Second Floor Plan

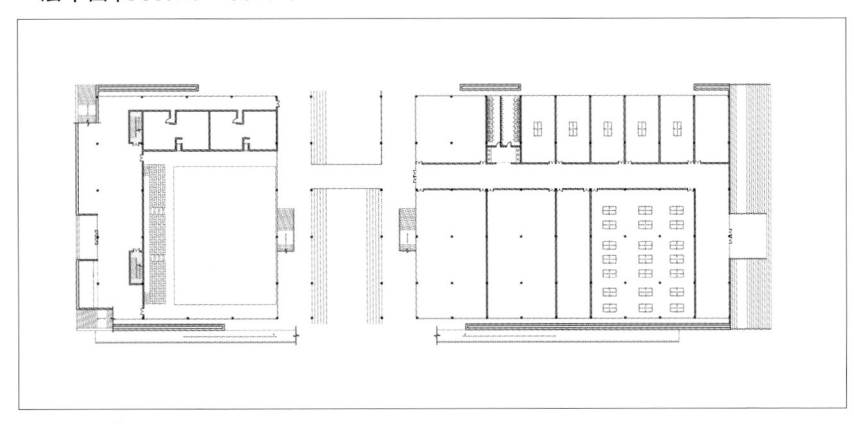

一层平面 | Ground Floor Plan

金陵体育
Http://www.jlsports.com

鄞州中学

Yinzhou Middle school

实 施 工 程 国际竞标中标 /2010—2015
方 案 图：高裕江、史国雷、郑颖生、毛志远、贾茜、许小笑、王何忆、饶峥、戴鹏杰、苏仁毅、王岳峰、丁思璐
施 工 图：高裕江、王敏霞、沈晓鸣、蒋君标、郭宁、赵鑫、饶峥、马云飞
施工图合作单位：宁波市城建设计研究院

采用相对集中，有机分散的手法，构筑新颖校园空间。顺应南北向“轴网形式”，构建校园主次轴线，展示理性、现代的特点。改造和充分利用内河道，营建富有特色的“绿在水中，水在绿中”的生态型校园格局。利用空间主轴与院落相融的空间模式，力求构筑尺度宜教、宜人的教学文化环境。

By organic combination, the project creates new campus space. The project follows the North-South grids to build the primary and secondary axis of the campus, and shows rational, modern features. By reconstruction and making full use of the river in the site, an ecological campus characterized by the "green in the water, the water in the green" is constructed. The space model of a combination of principal axis and courtyard is adopted, to build appropriate scale for teaching, teaching culture and pleasant environment.

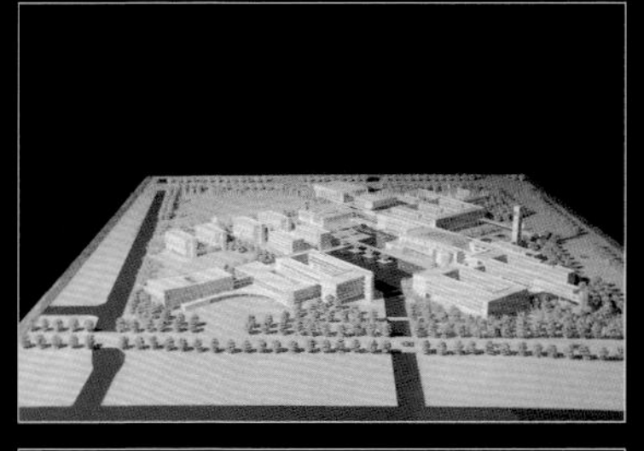

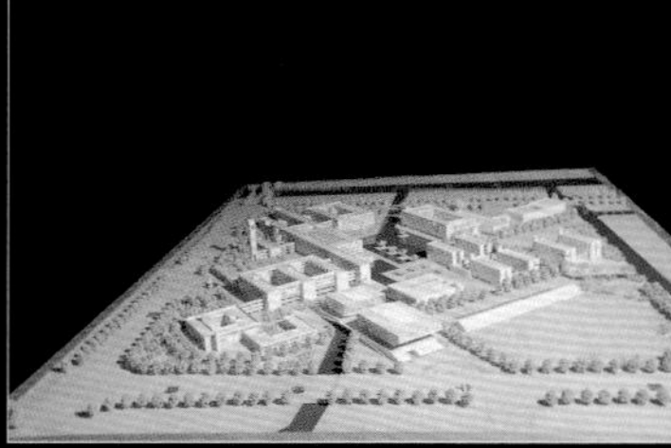

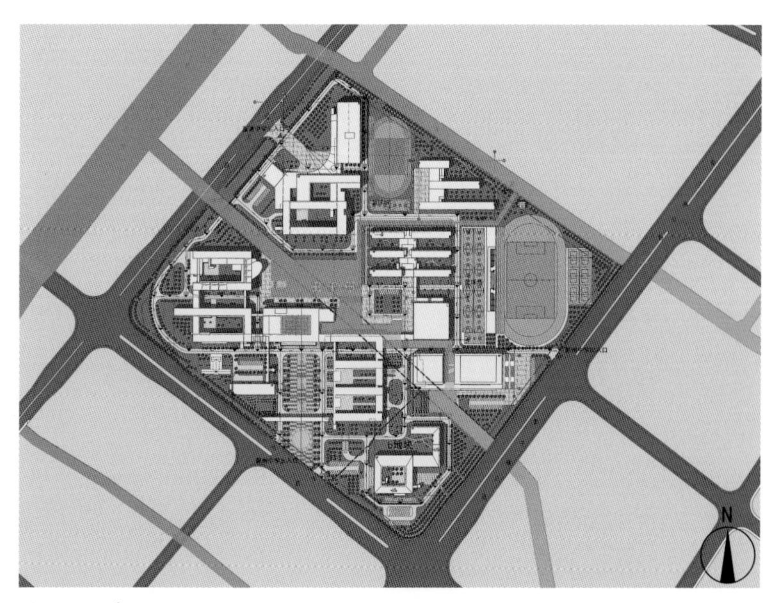

总平面图 | Site Plan

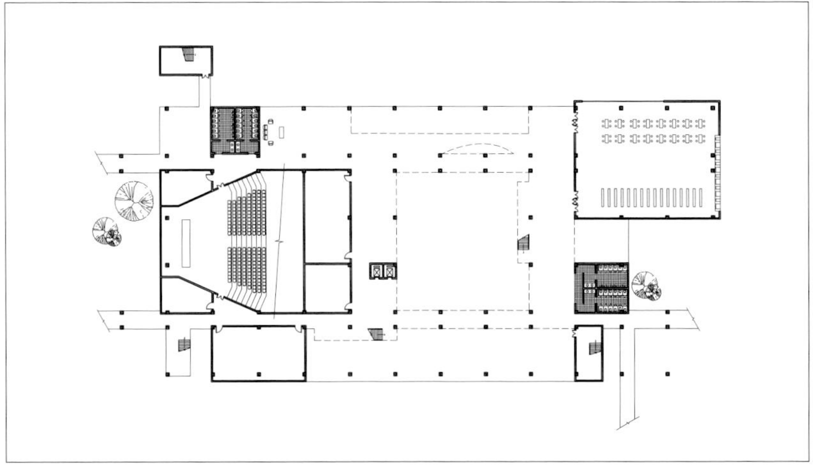

一层平面 | Ground Floor Plan

二层平面 | Second Floor Plan

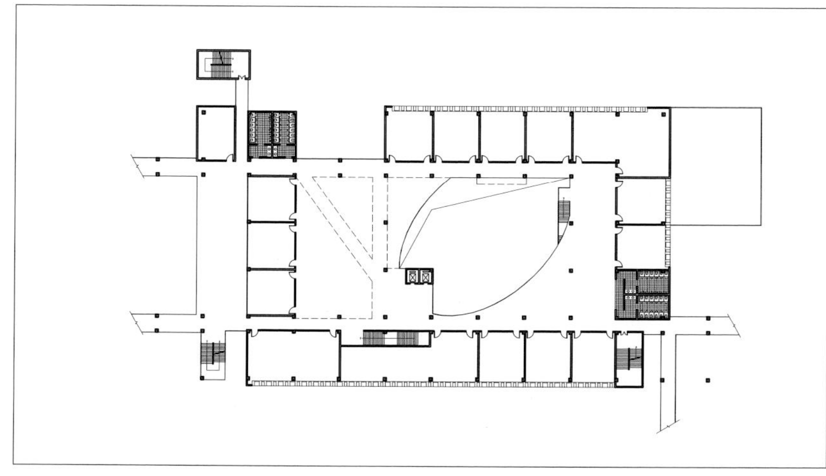

三层平面 | Third Floor Plan

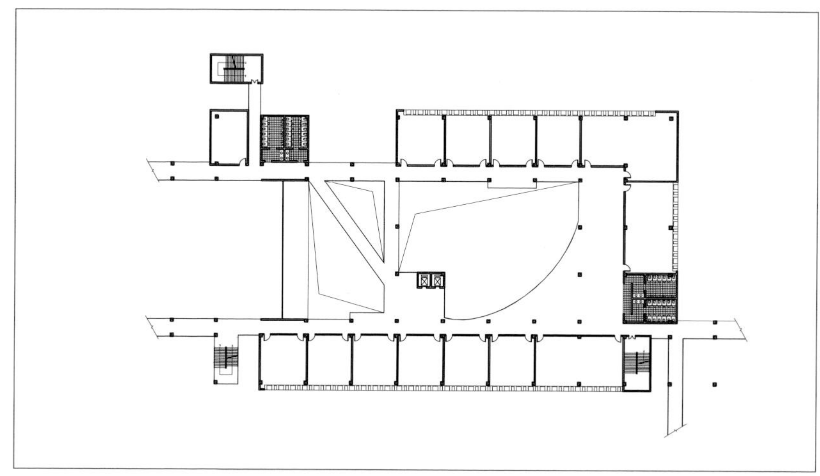

四层平面 | Fourth Floor Plan

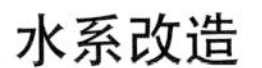

水系改造

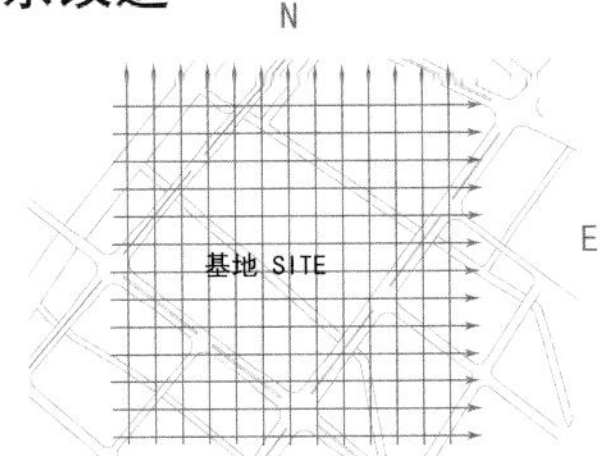

方向

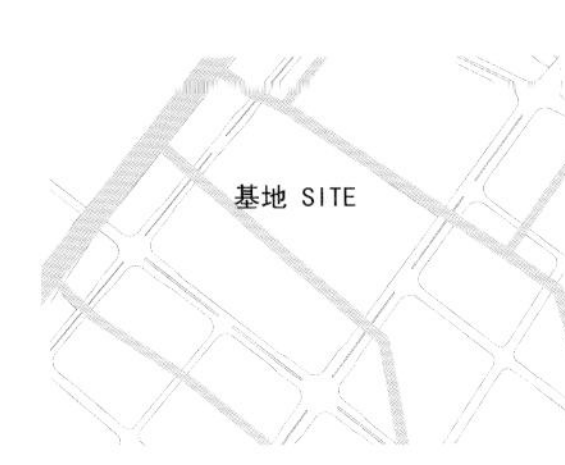

水网

路网

基地现状分析

原有水系

改造水系

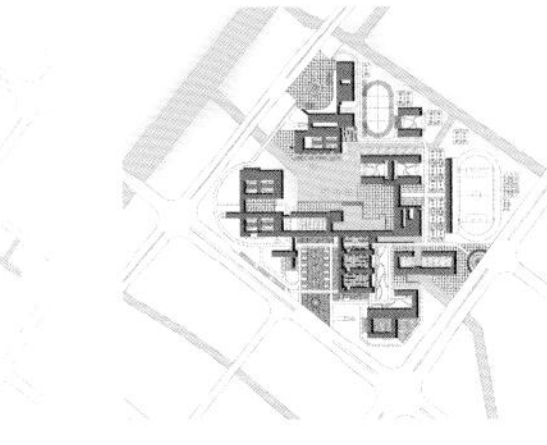

水系与建筑

水系改造分析

P
严禁车辆从
出口处驶入

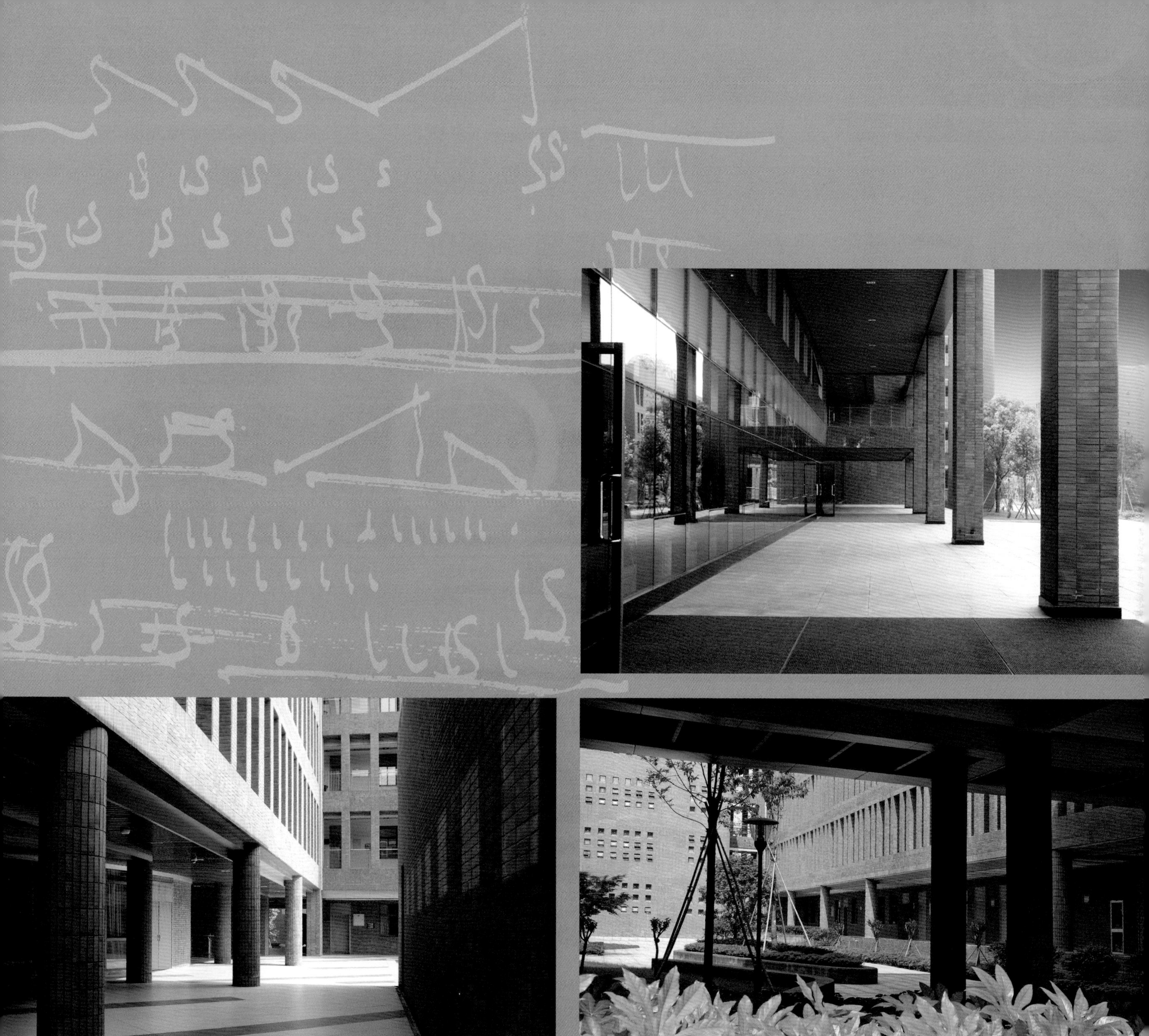

2016·鄞州中学优秀校友艺术作品邀请展
青出于蓝

巴西尹氏集团办公大楼

Brazil Yin's Group Office Building

实　施　方　案　国际中标 /2011
建 筑 师 团 队：高裕江、赵文玲、苏仁毅、丁思璐、
　　　　　　　　石碧军
施工图合作单位：宁波中鼎建筑设计研究院

源于“集装箱”叠合、整构而成的建筑形象，它是理性的，功能的，逻辑的；也是感性的，艺术的，文化的。设计表现为简约、现代的方正塔式形体，不仅满足高效、合理、可持续、生态的设计理念要求，而且隐喻“集装箱叠合”形态风格正展示了大楼主体——巴西尹氏的企业特色。

From the building image overlapped and integrated by "container",the architecture is rational,functional and logical.It is also emotional,artistic and cultural. The architecture is presented as a simple and modern square-tower form.It can not only meet requirements of efficient,reasonable,sustainable and ecological design concept,but also shows the characteristics of enterprises of the main building,namely Brazil Yin's through its form style from "container superimposed".

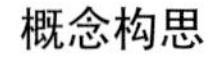
概念构思

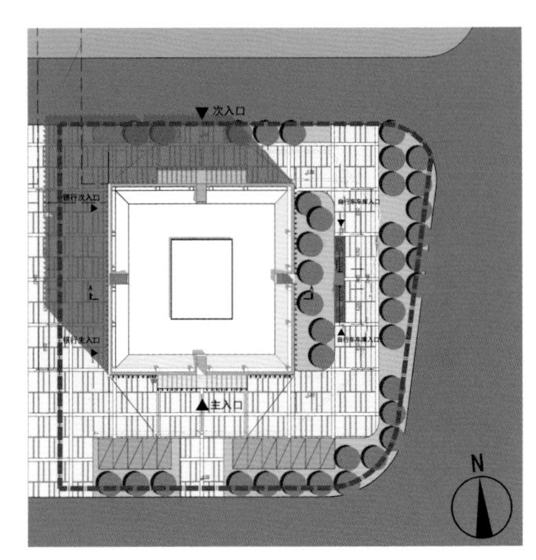

总平面图 | Site Plan

立面 | Elevation

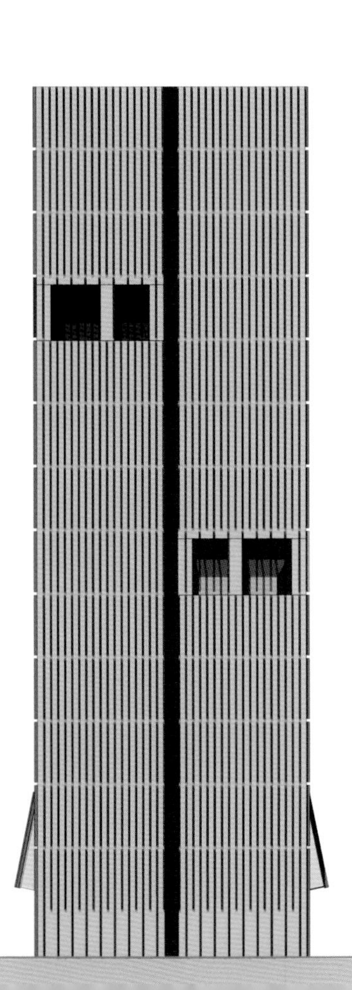

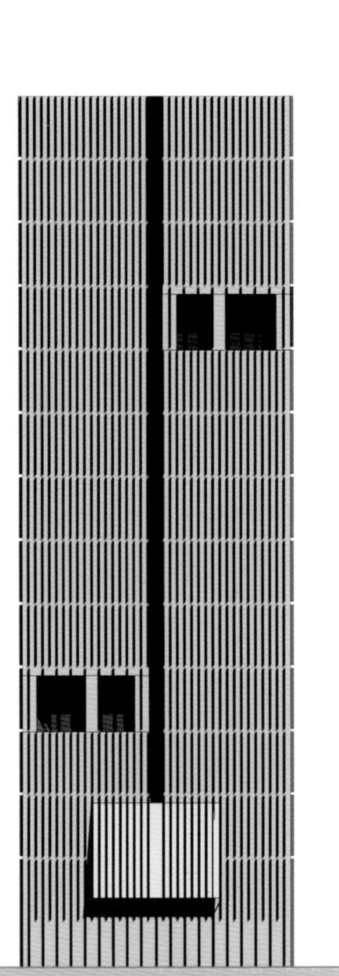

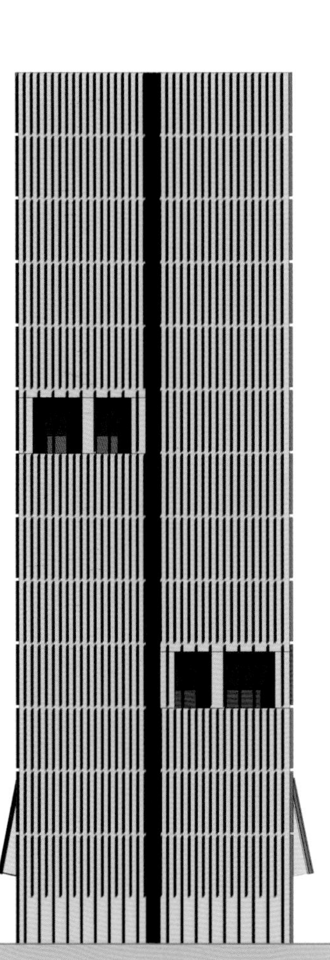

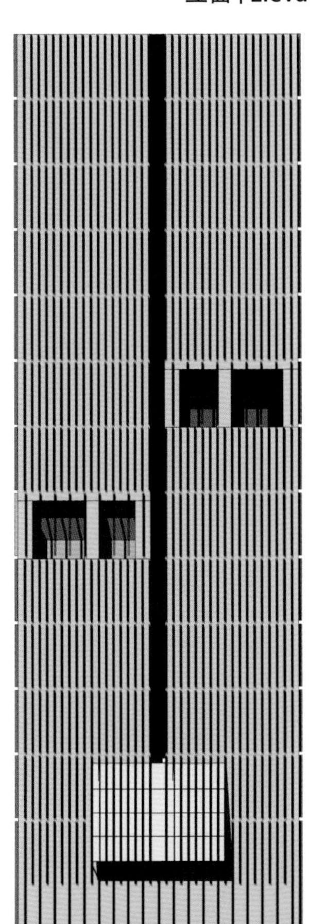

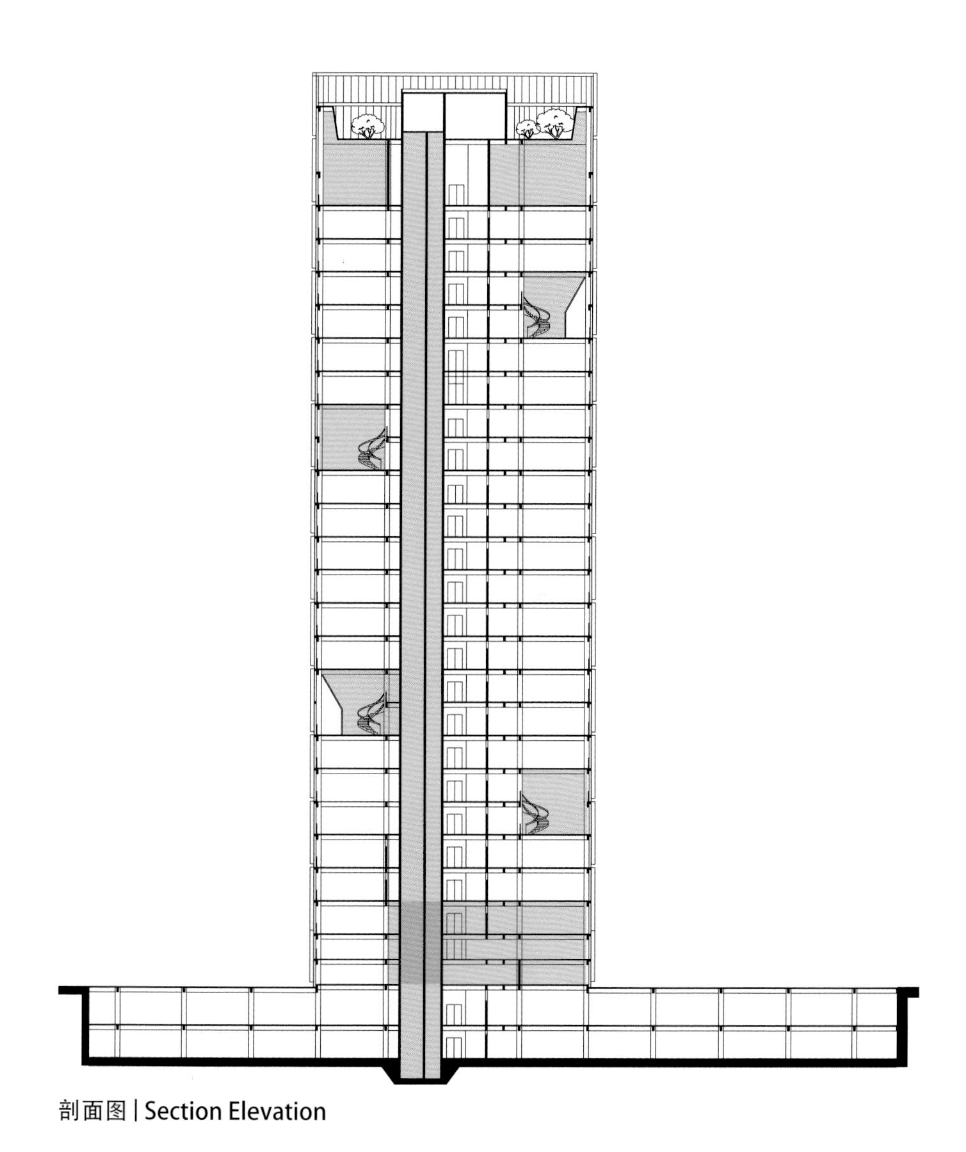

剖面图 | Section Elevation

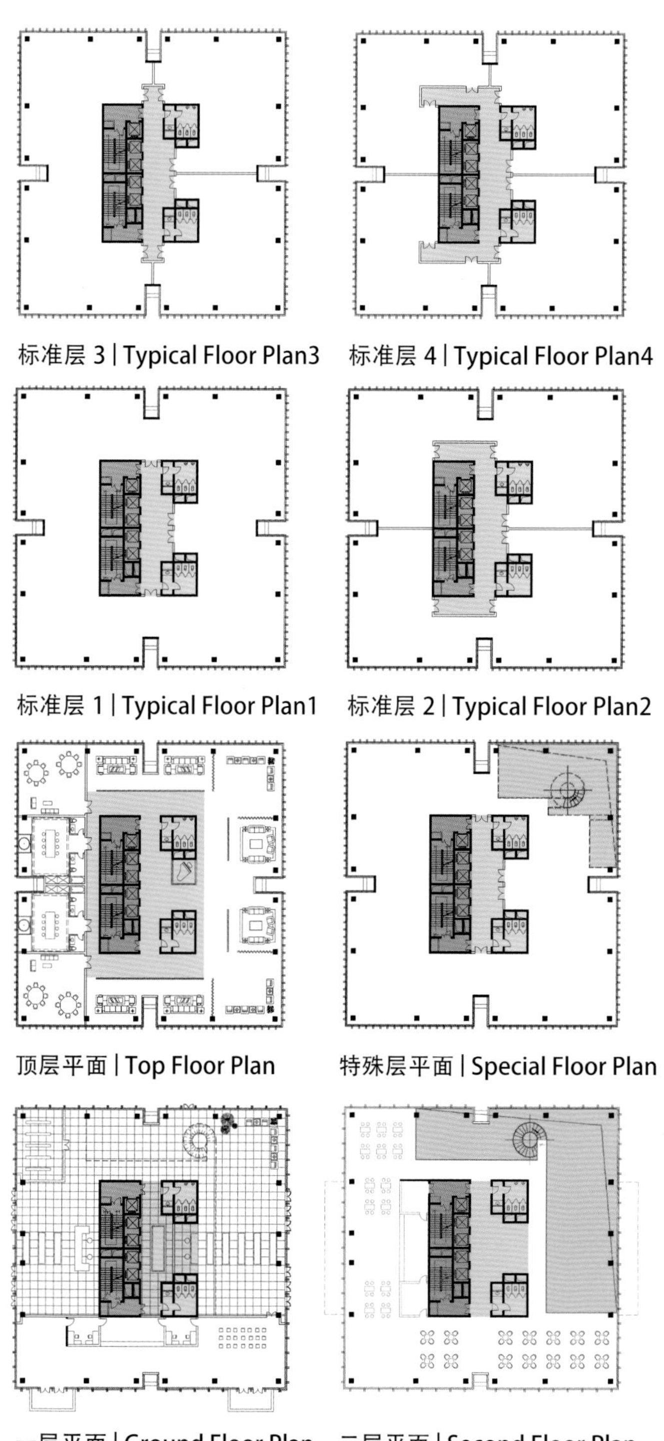

标准层 3 | Typical Floor Plan3

标准层 4 | Typical Floor Plan4

标准层 1 | Typical Floor Plan1

标准层 2 | Typical Floor Plan2

顶层平面 | Top Floor Plan

特殊层平面 | Special Floor Plan

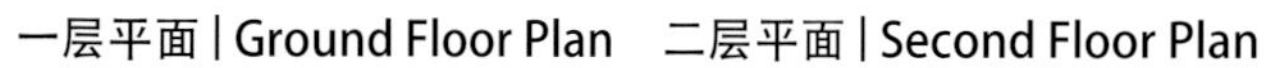

一层平面 | Ground Floor Plan

二层平面 | Second Floor Plan

宁波东钱湖千玉宫舍

Qianyu Museum of Ningbo Dongqian Lake

概 念 方 案　国际竞标 /2011
建筑师团队：高裕江、史国雷、饶峥、王何忆、苏仁毅、王岳峰

设计以吴越春秋时期宫城、台基为原型，整合秦汉“四坡顶”的建筑元素，形成“似城非城，似屋非屋”的建筑形态风格。以传统宫城轴线串联起“水院”“宫舍”“内院”“里院”等不同文化生活主题的院落型空间，丰富空间层次，烘托文化氛围，强化尊贵与皇家之气。

Design is based on the prototype of imperial palace and pedestal in Wuyue Chunqiu period, integrating the Qin and Han "hipped roof" as architectural elements, forming a "non-city like the city, like house non-house" style of architectural form. The traditional axes of imperial palace are adopted to put together different cultural courtyard type space as "Water Yard" "Palace House" "inner courtyard" "courtyard", enrich space levels, heighten culture, strengthen the noble and royal atmosphere.

草图 | Sketch

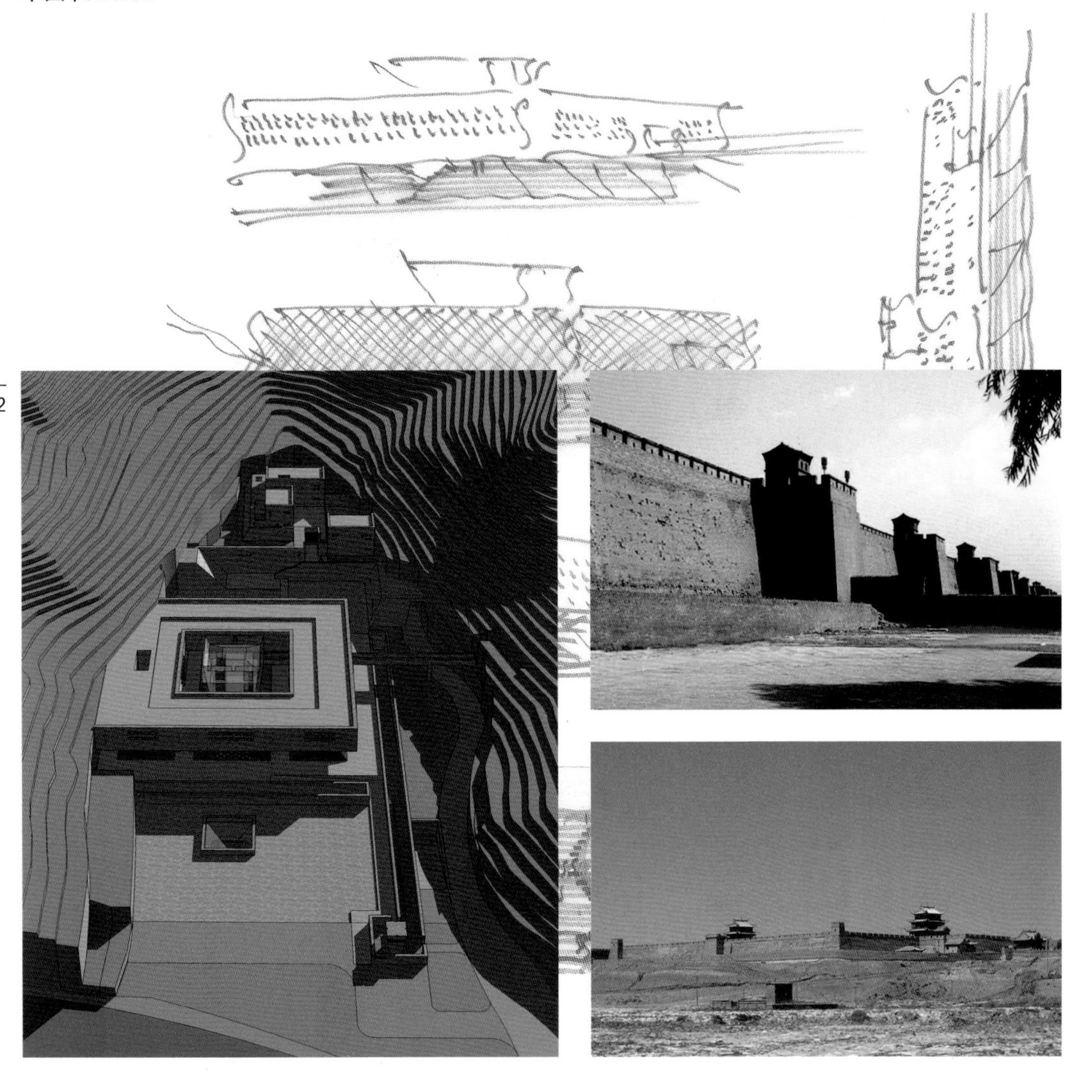

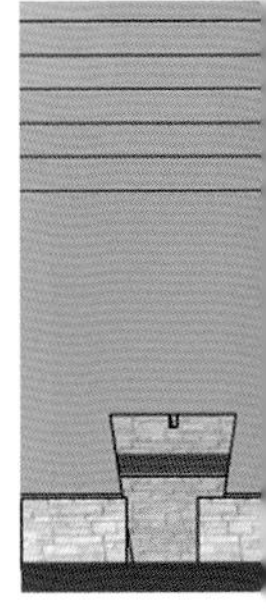

北立面 | North Elevation

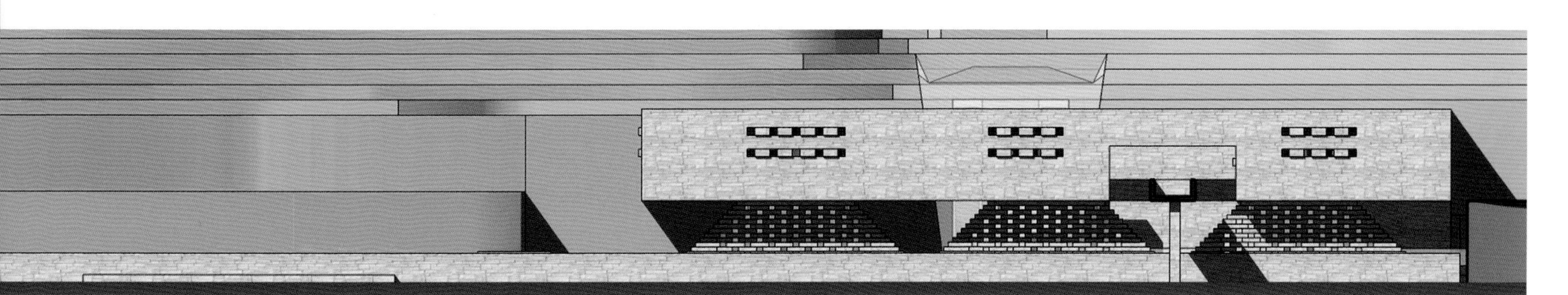

立面 | West Elevation

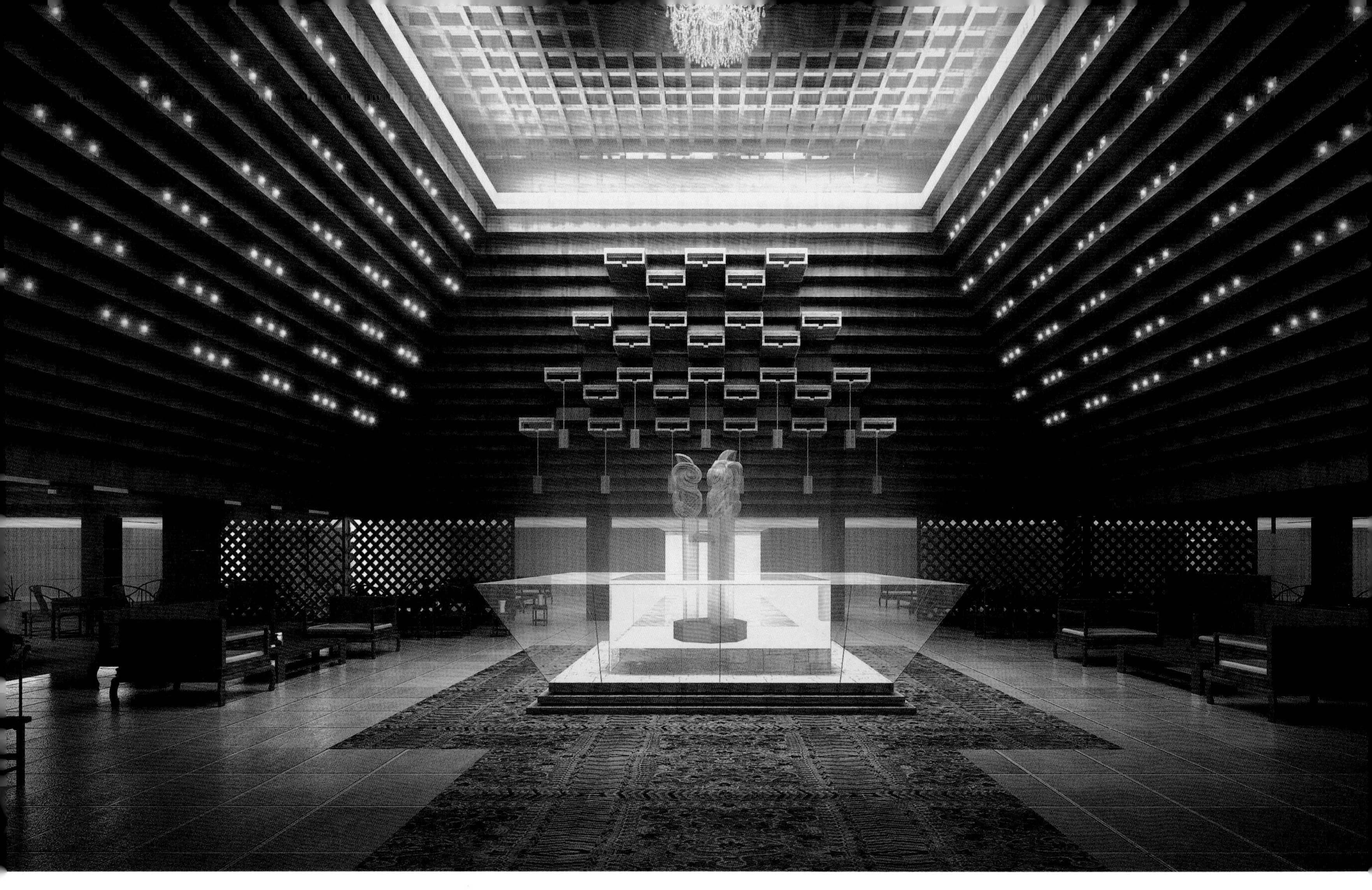

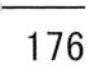

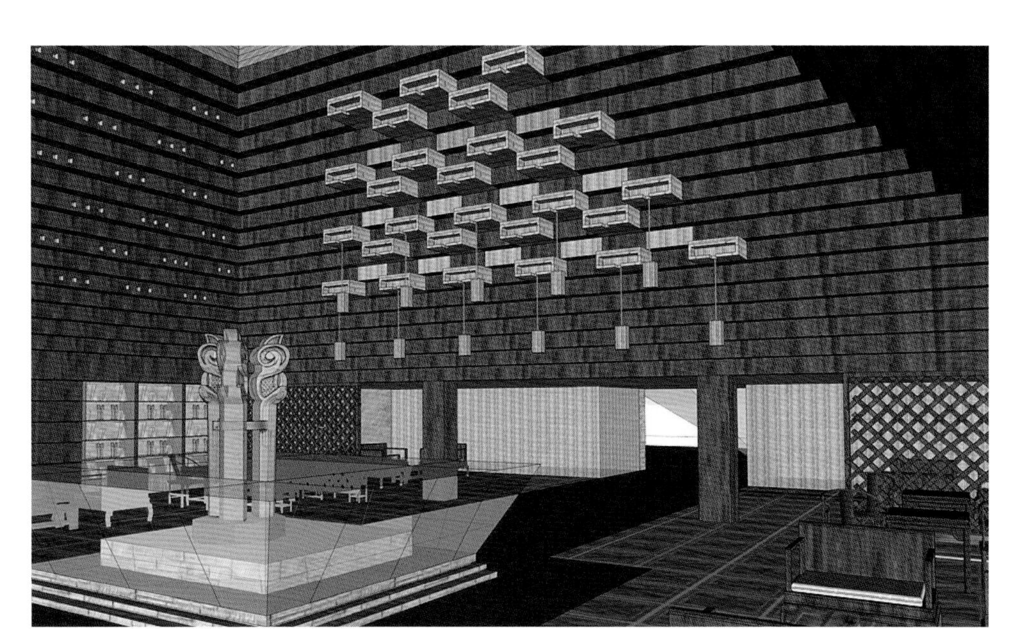

方案比较

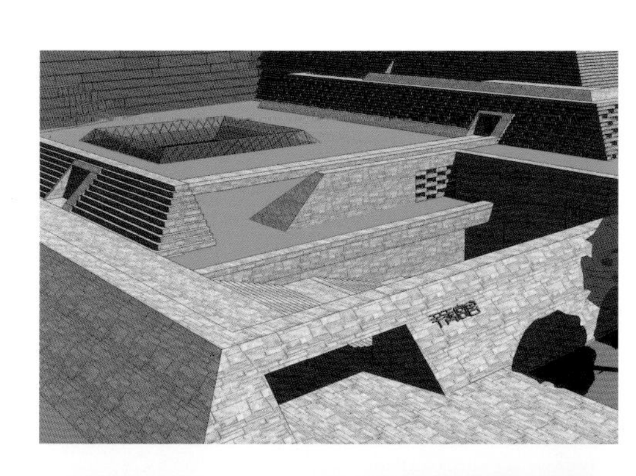

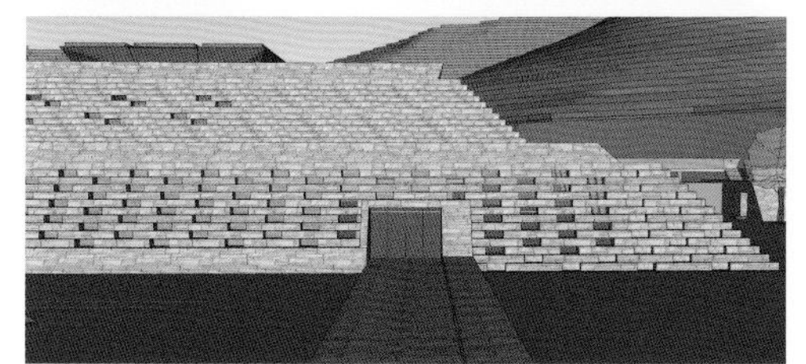

浙江大学舟山校区

Zhoushan Campus of Zhejiang University

中标规划方案 2012
建 筑 师 团 队：高裕江、王何忆、屠芳奇、史国雷、郑颖生、饶峥、杨晋、王豪、马广川、戴鹏杰

如何充分体现"依山融湖，凭海为邻"基地自然条件，突出"山水"要素，强化"海洋"特色，"使新校区不仅成为一所现代气息浓厚的大学校园，更要成为浙江大学及舟山国家级新区对外开放的一张靓丽名片"，这是规划设计的第一要点。

规划设计如何鉴于群山相夹不甚规则的用地现状，以及新城海天景观大道的城市建设要求，并能借湖水及滨海特色风光之景，使城市滨海景观与高校校区的景观环境相整合，这也是规划设计的另一个关键点。规划设计如何既能运用东方传统"背山面水"的规划理念，又能充分整构山湖及城市开放空间，符合节能、节地、节水等绿色生态的理念要求，这是本规划设计的又一要点。建筑形态设计如何传承浙江大学校园建筑文化，而能突出舟山校区海洋学院应有特色，同时又可避免近海大风、大浪等风雨气候的不利影响，是建筑空间形态设计必须面对的又一主题。

There are four challenges of this project.The biggest challenge is to demonstrate the beautiful natural environment of HuiMing Lake, HuiMing Mountain and the sea nearby, making the Zhoushan Campus be a modern University campus as well as an identity landscape of the Zhejiang University.The second challenge is about how to take the advantage of the irregular site to satisfy the demand of the urban construction and create a campus landscape coping with the sea side view of the city.The third challenge is to use the traditional Chinese planning concept of "Fronting water and with hills on the back" as a leading concept to create an energy-saving, land-saving and water-saving campus.The last challenge is to highlight the architectural character of this site and create safe and sound spaces to resist the big wind and sea waves may caused by the sea nearby.

草图 | Sketch

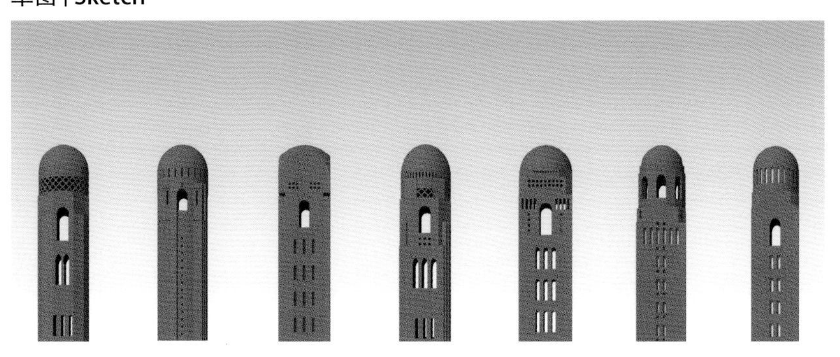

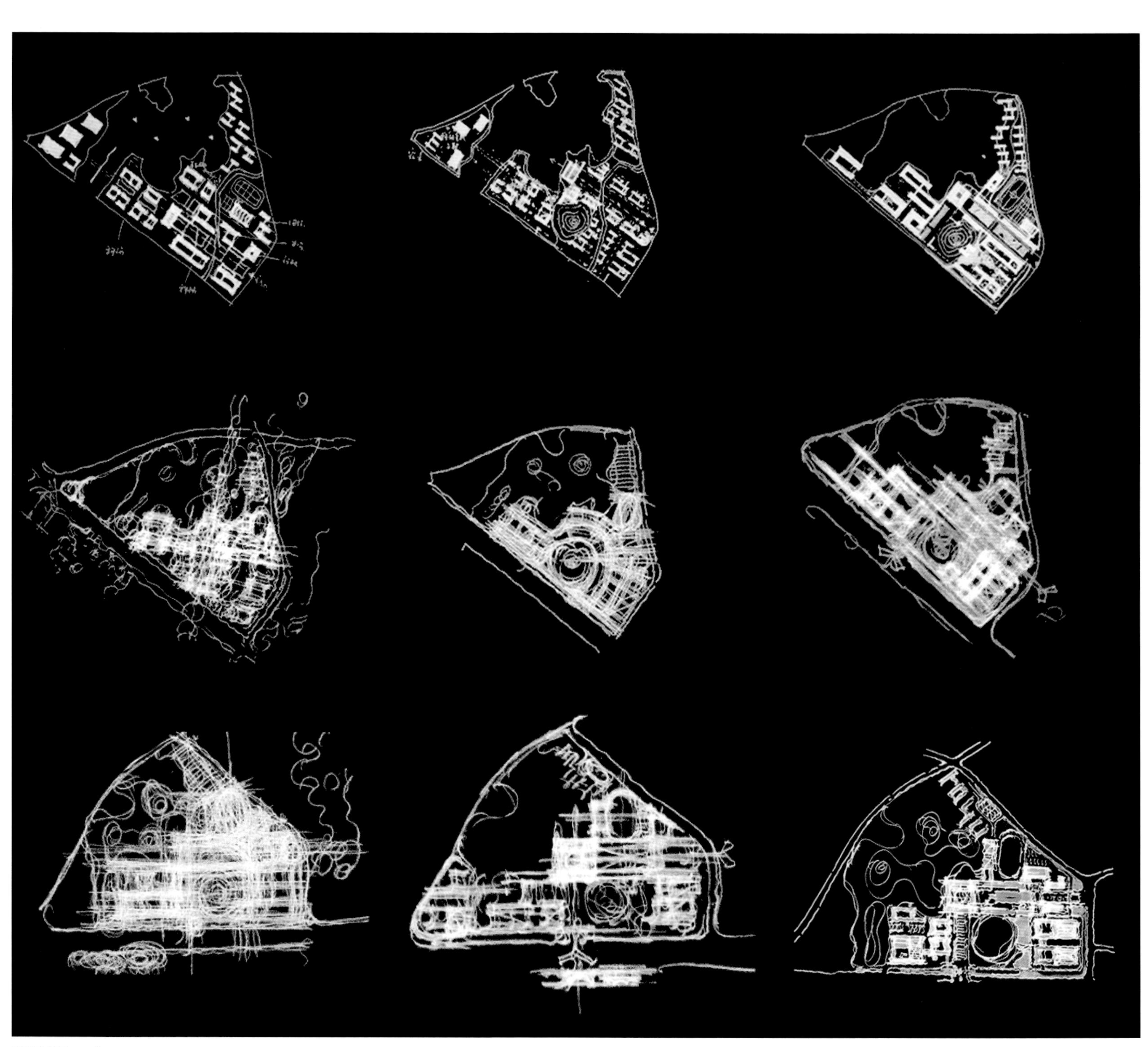

草图 | Sketch

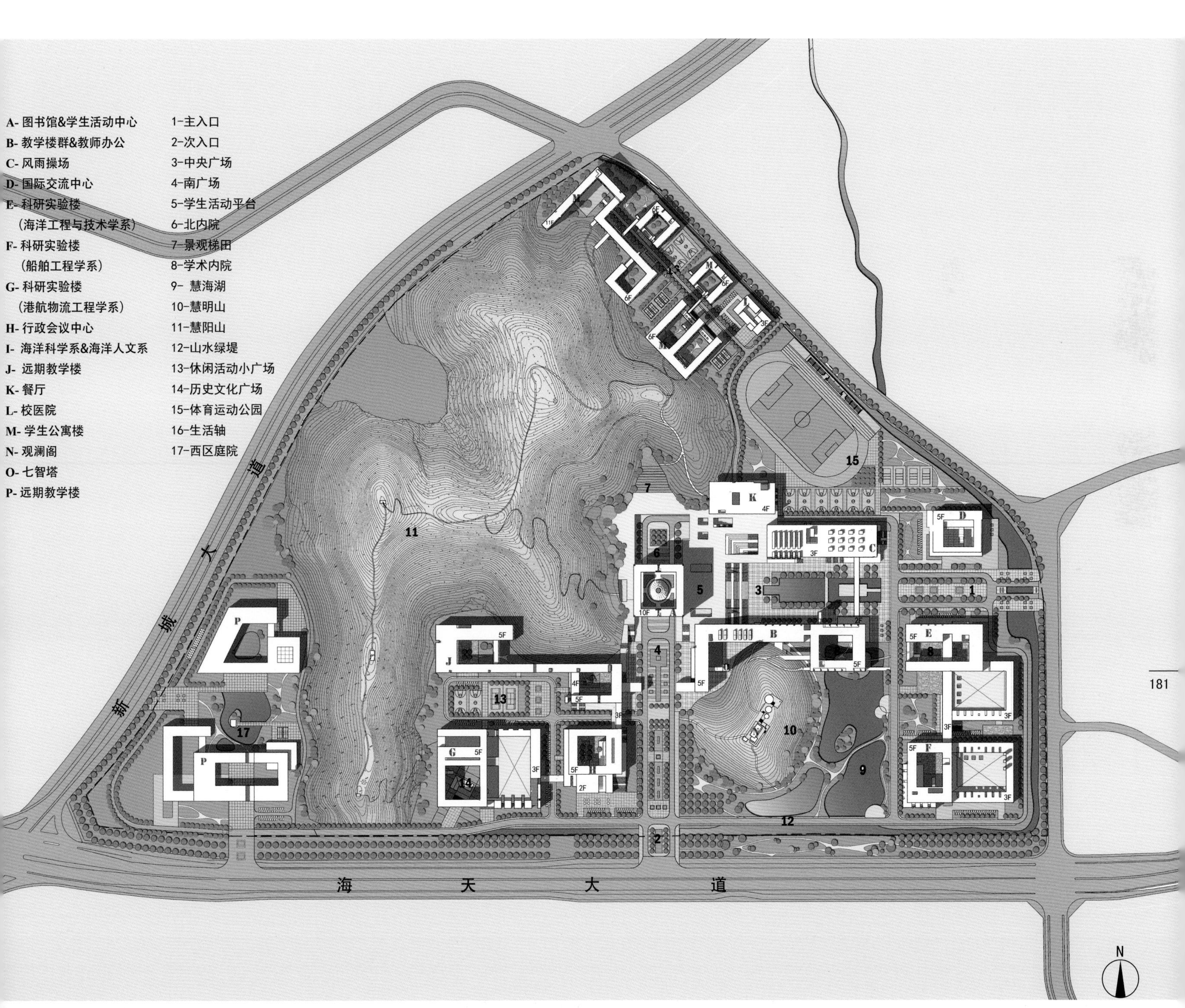
A- 图书馆&学生活动中心
B- 教学楼群&教师办公
C- 风雨操场
D- 国际交流中心
E- 科研实验楼
（海洋工程与技术学系）
F- 科研实验楼
（船舶工程学系）
G- 科研实验楼
（港航物流工程学系）
H- 行政会议中心
I- 海洋科学系&海洋人文系
J- 远期教学楼
K- 餐厅
L- 校医院
M- 学生公寓楼
N- 观澜阁
O- 七智塔
P- 远期教学楼
1-主入口
2-次入口
3-中央广场
4-南广场
5-学生活动平台
6-北内院
7-景观梯田
8-学术内院
9- 慧海湖
10-慧明山
11-慧阳山
12-山水绿堤
13-休闲活动小广场
14-历史文化广场
15-体育运动公园
16-生活轴
17-西区庭院
新
城
大
道
海
天
大
道
N

-面图 | Site Plan

平改坡风格比较

宁波钟公庙综合楼

Business and Government Office Complex in Ningbo

概 念 方 案　2013
建筑师团队：高裕江、屠芳奇、王何忆、马云飞、何雅俊

设计采用整体构架的形态模式，来应对周边不同的建筑形态。通过中央庭院（约 75 米×65 米）的设置，以改善不利的环境景观，有效提高中央庭院景观空间的共享度。建筑设计运用江南传统民居格栅和窗棂的变异形式，结合隔热保温真空幕墙玻璃和遮阳构件，承延地域的建筑文化脉络，展示出建筑文化性和艺术性。

To cope with multiple architectural form nearby, the designers adopt a mega-structure and a center courtyard to perfect the adverse environment nearby and increase the sharing extent of the garden view. Besides, the variant window lattice of the traditional Chinese dwellings combined with the heat insulation glass walls and sun shading constructions are used to follow the traditional architectural context as a demonstration of the artistry in this project.

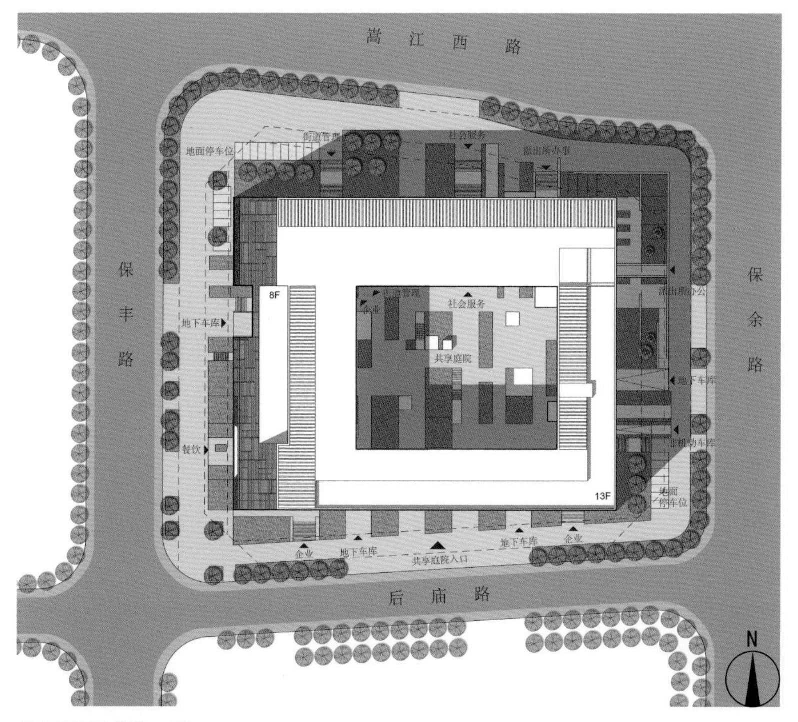

总平面图 | Site Plan

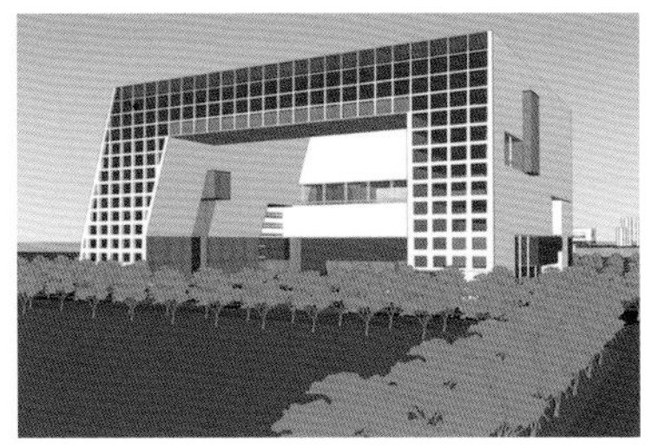

立面推敲

北立面 | North Elevation

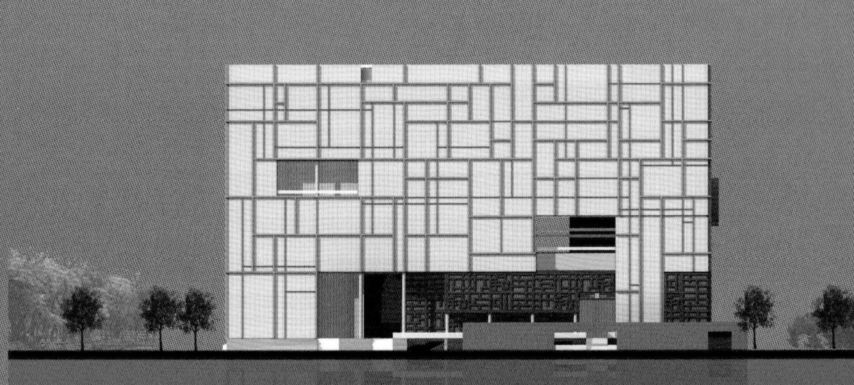

东立面 | East Elevation

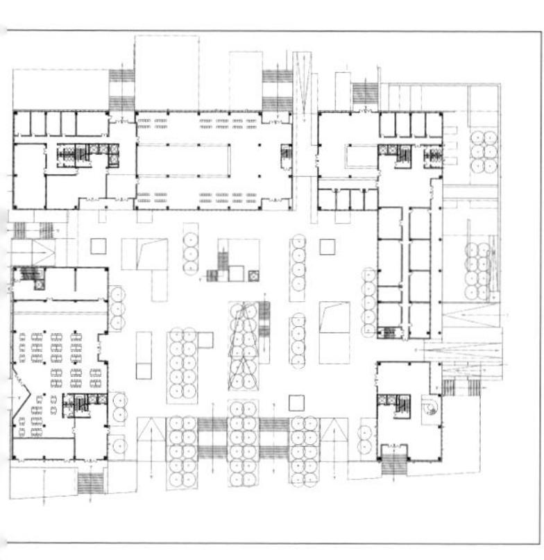
层平面 | Ground Floor Plan

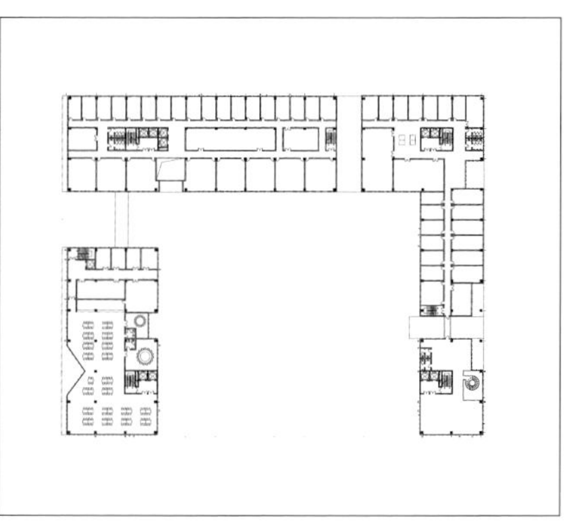
二层平面 | Second Floor Plan

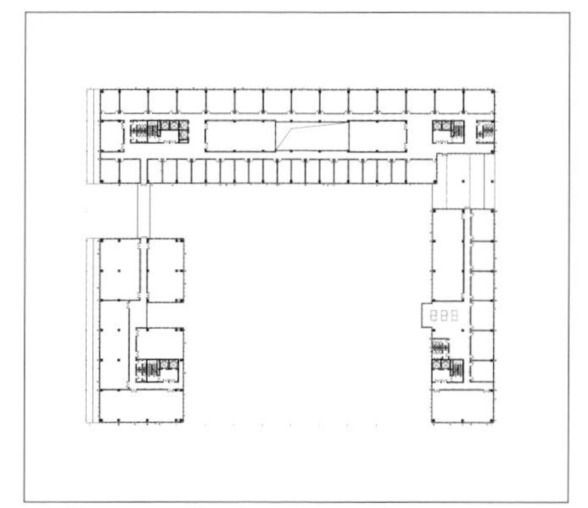
五层平面 | Fifth Floor Plan

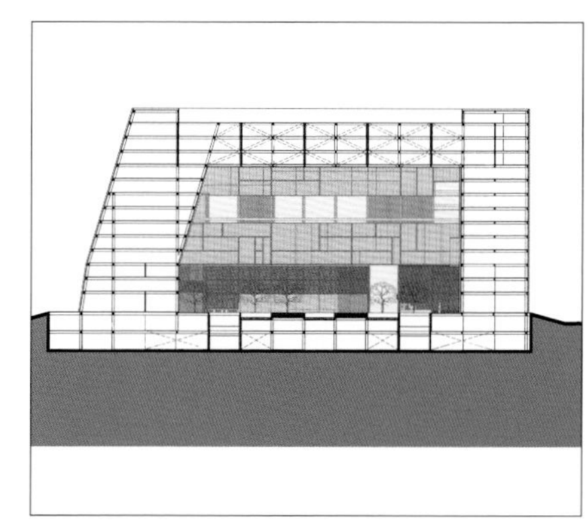
剖面图 | Section Elevation

温州苍南灵溪文化中心

Cultural Center in Lingxi Town, Cangnan, Wenzhou

在建项目　2013—
方 案 图：高裕江、何雅俊、周璐瑶、马云飞、饶铮、季群珊、王晓帆、徐崭青、邓奥博
施 工 图：高裕江、鲁丹、胡冀现、周璐瑶、邓奥博、苏仁毅、杨都、沈磊

设计采用“整构一体”策略，充分尊重周边环境，结合基地湖景，运用生态、地景建筑设计理念，既分又合，隔而又联。运用几何理性手法，使建筑内在功能和形象特色有机融汇，再现江南传统庭院，传承江南民居的形态模式。使用灰色石材、光洁金属构件与玻璃组构，演绎出具有时代气息和地域特性的建筑艺术气质。

This design adopts " integration" strategy, with full respect for the surrounding environment, combined with the base of the lake, the conception of ecological and landscape design, separated but linked. Using geometric rational skills, make the building an organic fusion of internal functions and image characteristics. Reproduce the southern traditional courtyard to inherit the southern residential shape pattern. Using gray stone, smooth metal components and glass fabric to illustrate an architectural temperament of the times and geographical characteristics.

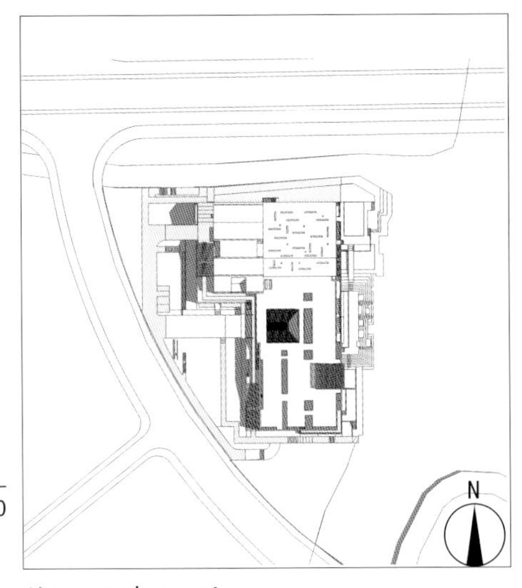

总平面图 | Site Plan

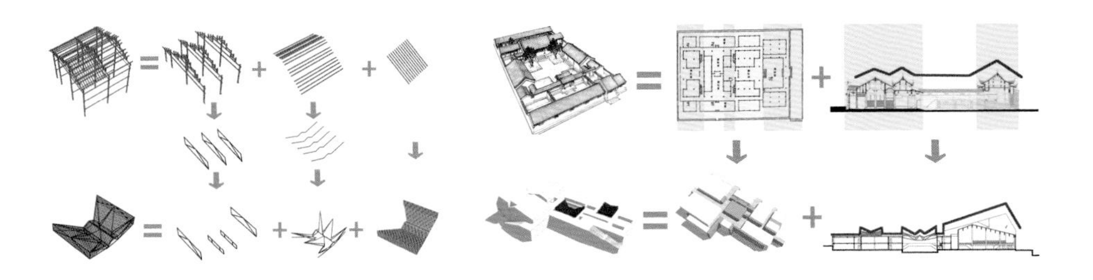

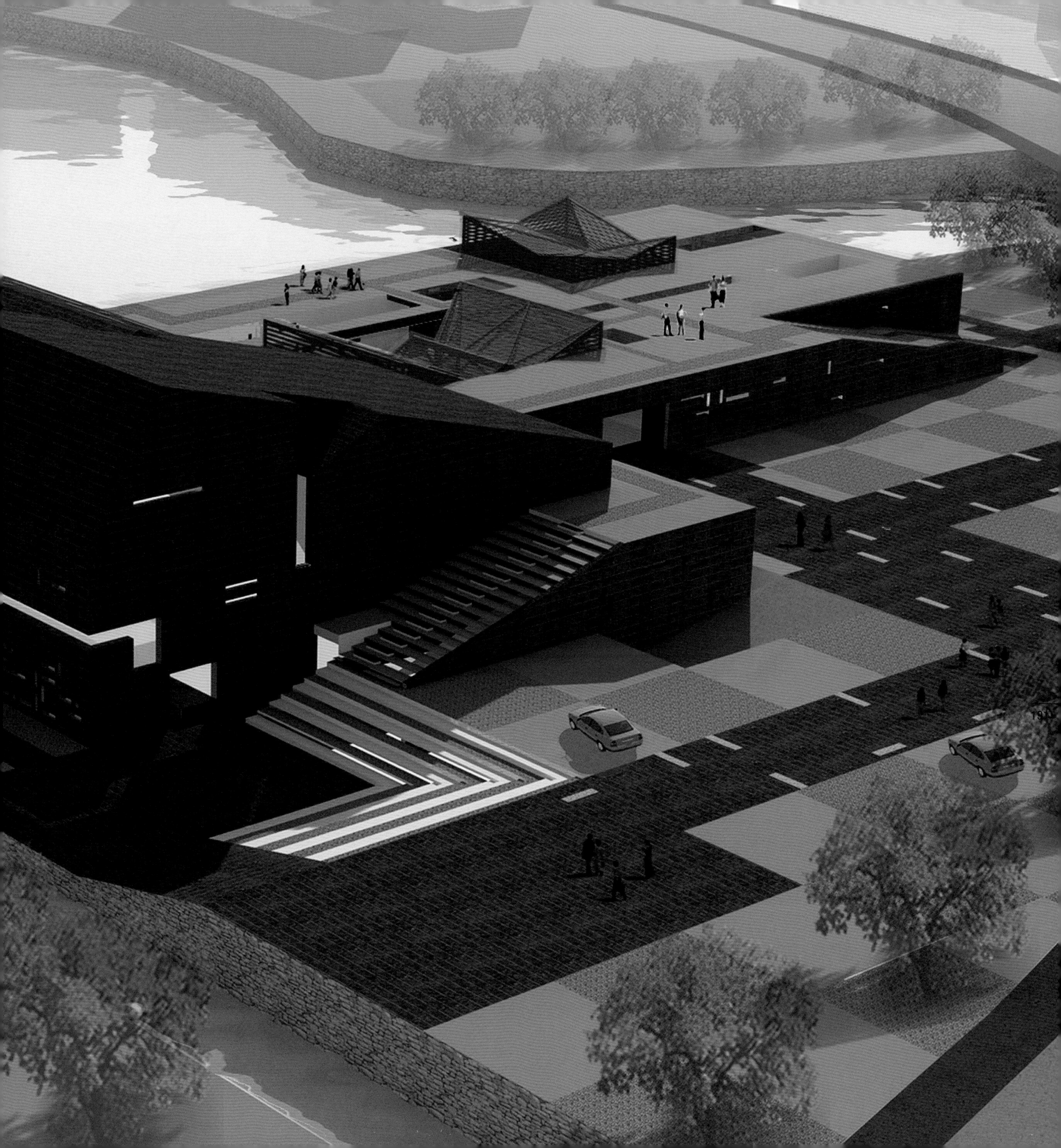

概念方案

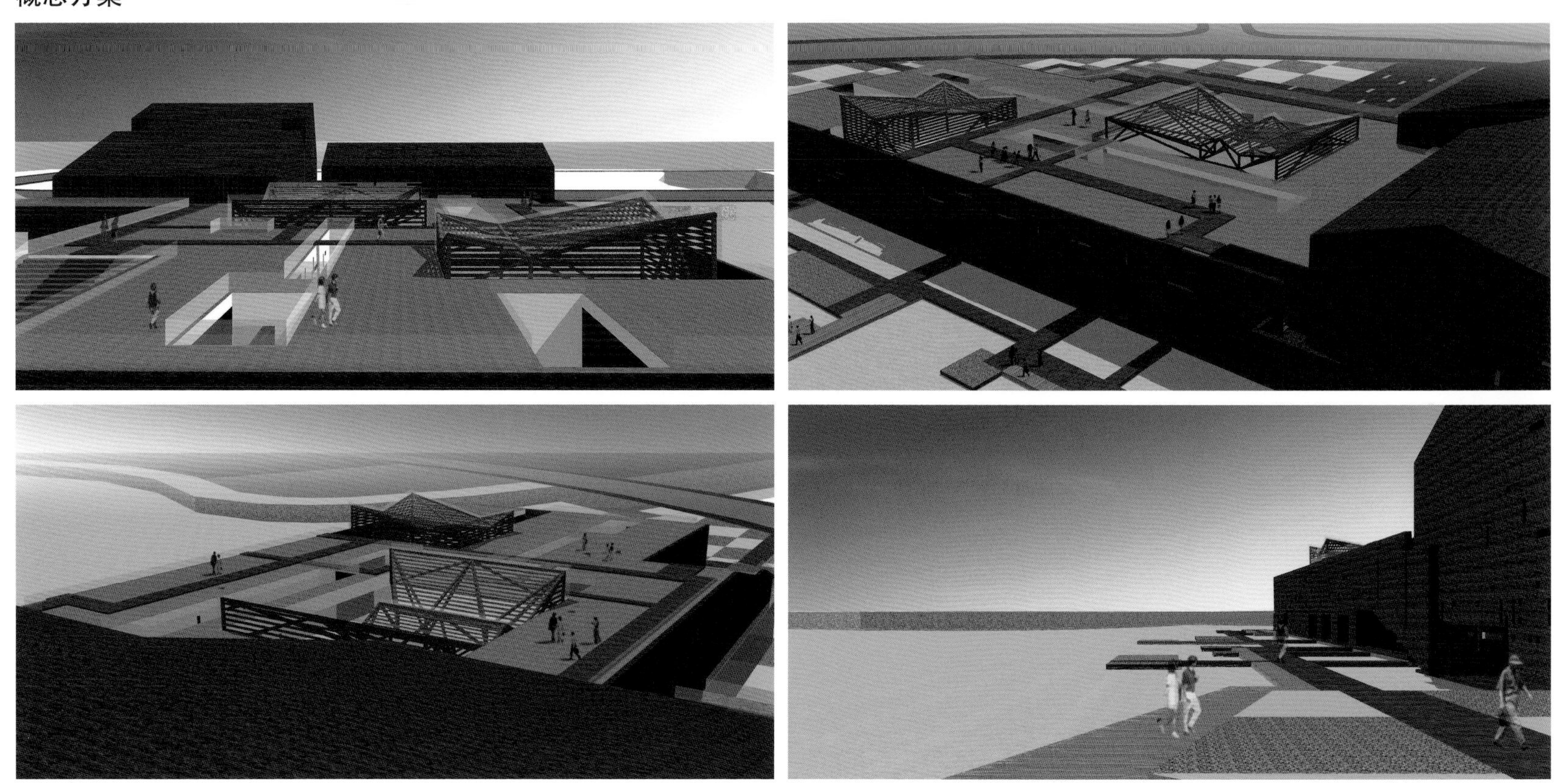

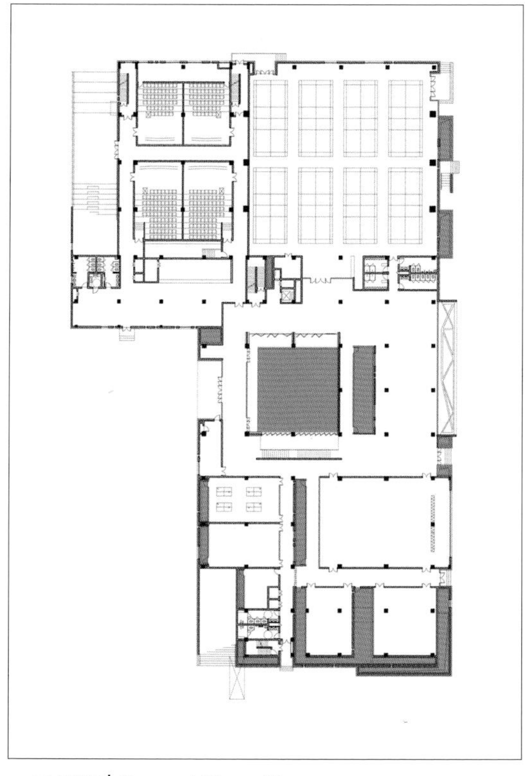

一层平面 | Ground Floor Plan

二层平面 | Second Floor Plan

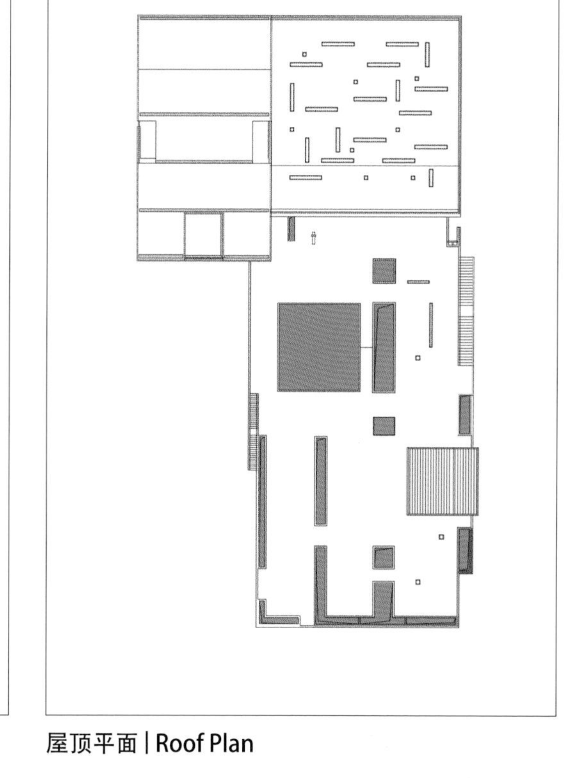

屋顶平面 | Roof Plan

深化方案

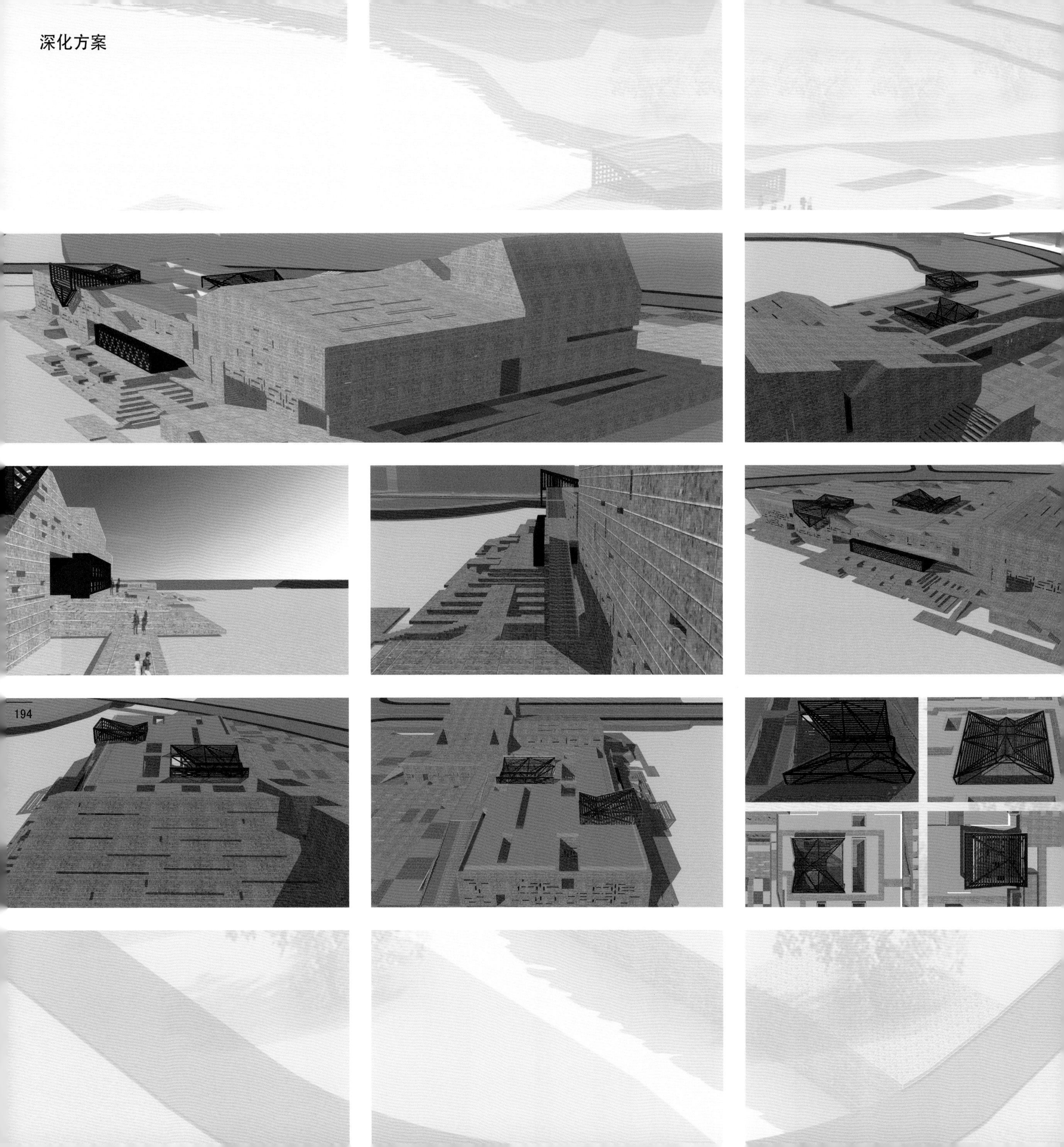

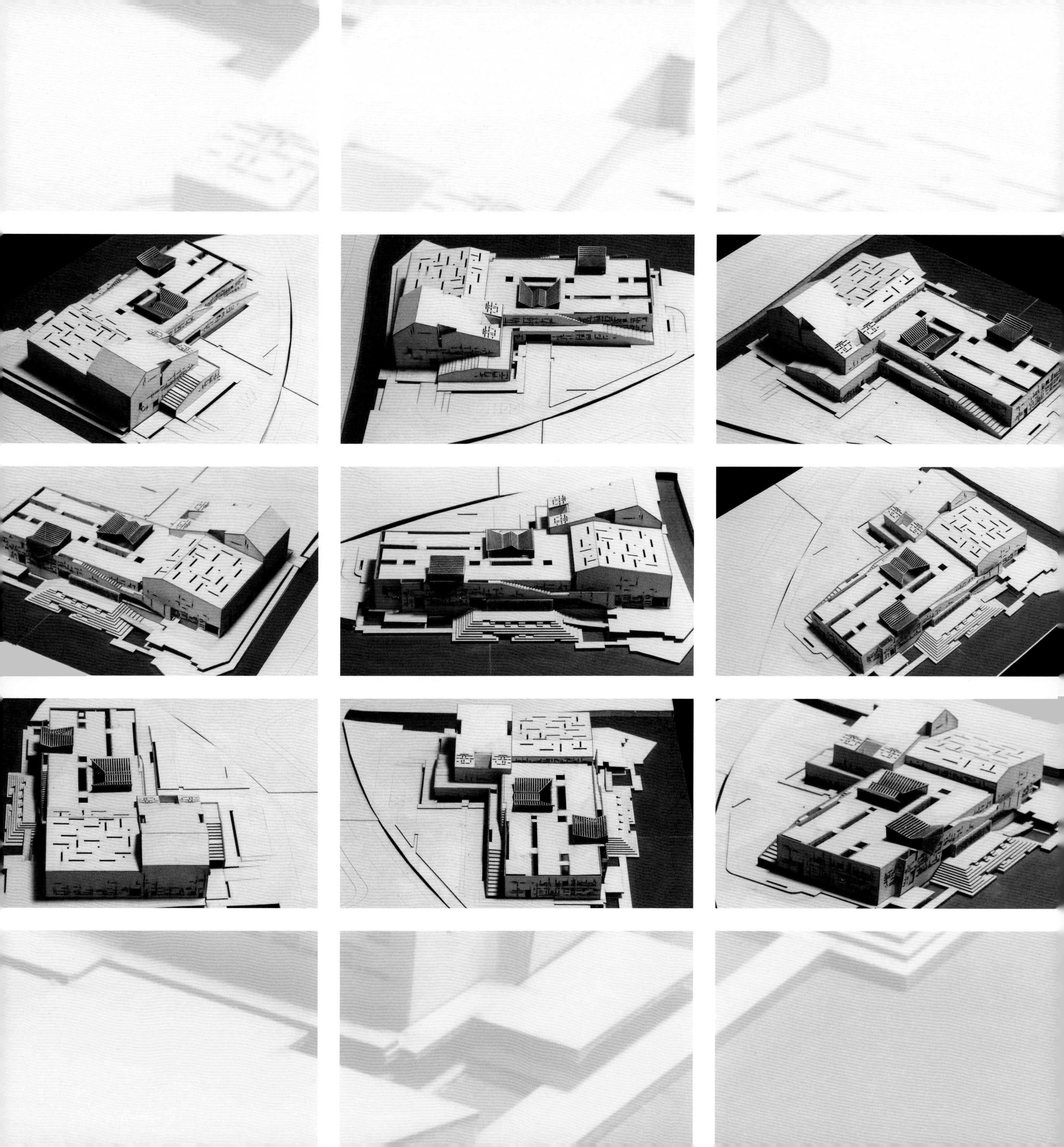

阅览室

COFFEE BAR

方案比较

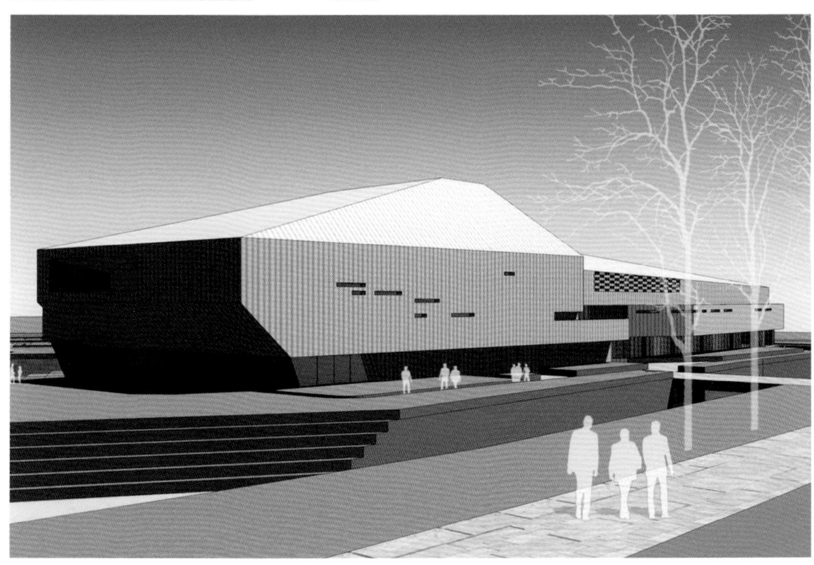
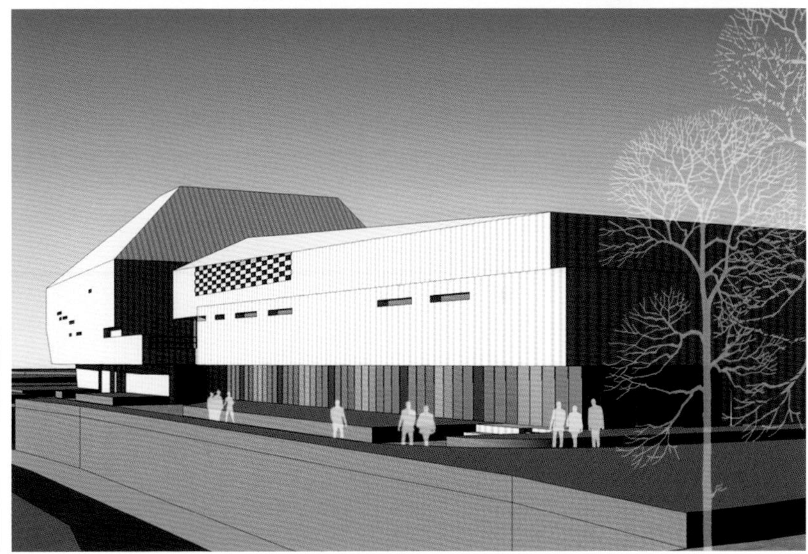

浙江抗日战争胜利纪念馆

Zhejiang Memory Hall of the Victory of the Anti-Japanese War

实 施 工 程　2014—2015
建筑师团队：高裕江、董志平、徐崭青、季群珊、王晓帆、喻丹

建筑利用原有场地落差，以原受降厅地坪为基准，营造纪念性场所景观，并将主体陈展功能置于场地之下，寓意灾难深重的近现代中华民族。在新构建的场所形成阶梯广场，在其上，创造“V”字形的形体构架，置于基地的西北侧，与“场所”形成同构关系，昭示抗战的最终胜利。纪念馆建筑的形象以突出“场所”与“氛围”为主，弱化常规的建筑形象，以石料为装饰主体，展示庄重永久的建筑场所。

Located on the lowered excavated part near the old memorial hall, the new memorial hall is designed as a half-buried building thus to create a memorial landscape that connected to the old building. The underground main building is a metaphor for the suffering Chinese people had been through the war, and the "v" shaped construction above reflects the victory of Anti-Japanese war. From the main entrance, there is a winding pathway that descends into the main exhibition hall. The new memorial hall is in a sense an organic architecture that integrates into the nature landscape as an extension of the old memorial hall, which highlighting the “atmosphere" and weakening the architectural image. The whole building is decorated with stone, thus make the memorial hall in a sense of eternity.

草图 | Sketch

5.9.4—9.5
宋殿受降
浙江受降

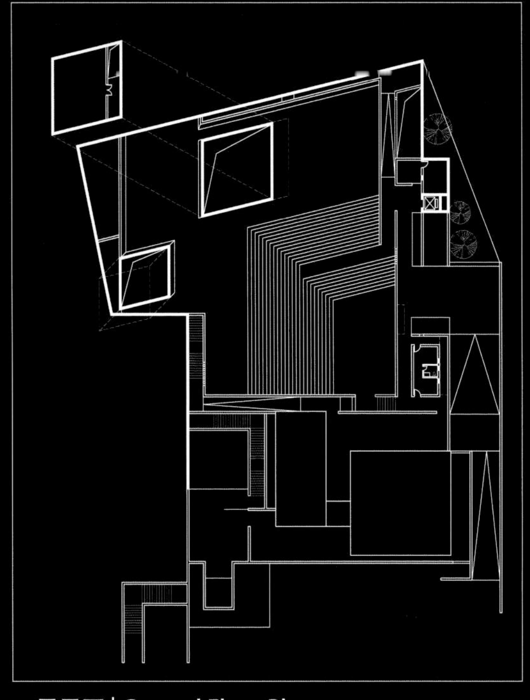

一层平面 | Ground Floor Plan

地下夹层平面 | Underground Mezzanine Floor Plan

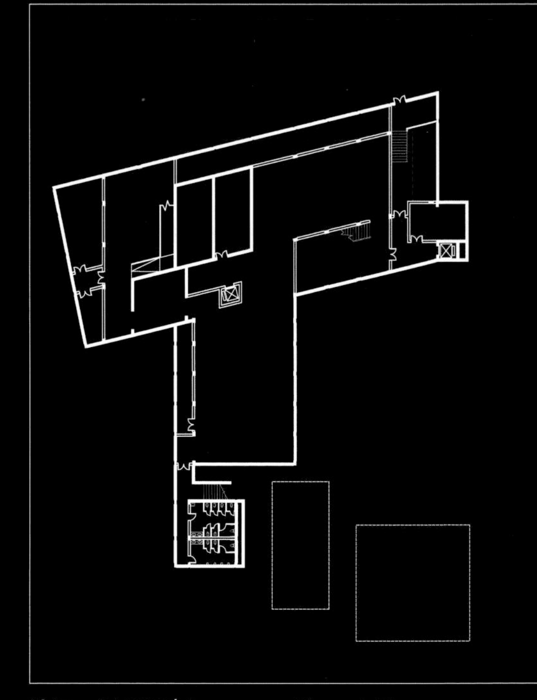

地下一层平面 | Basement Floor 1 Plan

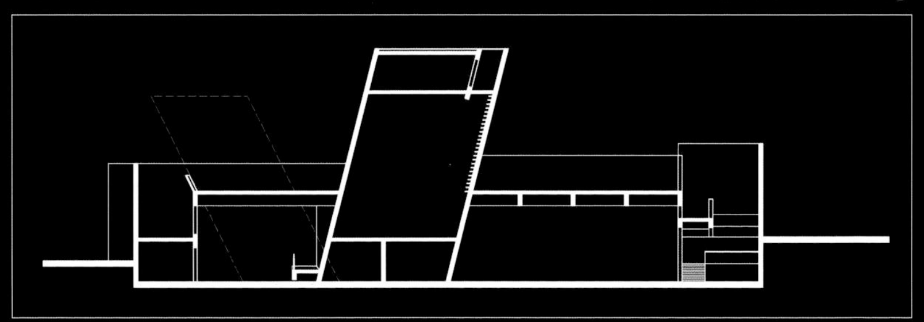

剖面图 | Section Elevation

剖面图 | Section Elevation

东南立面 | South-East Elevation

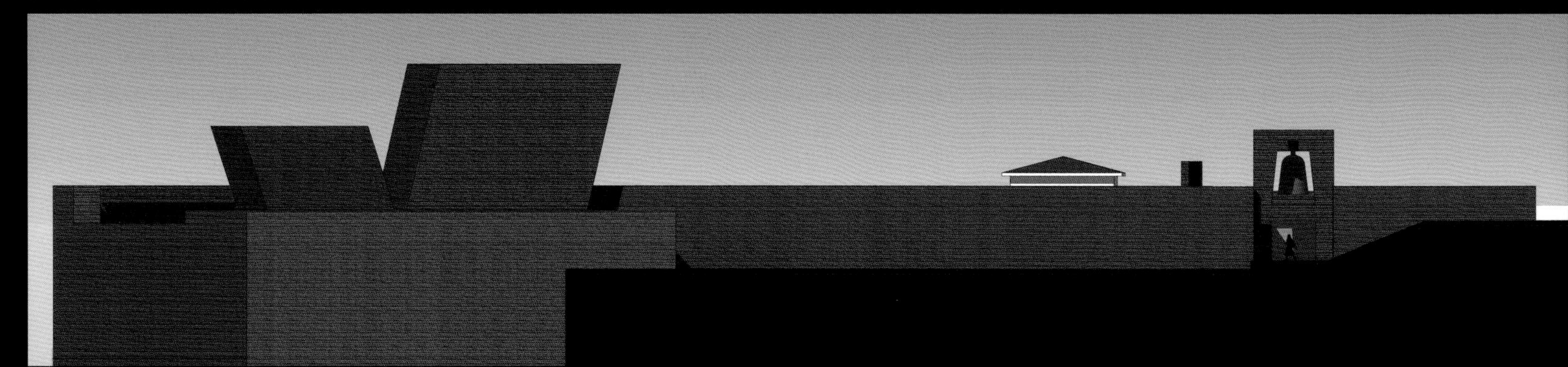

后 记

《中庸建构——浙江大学建筑系高裕江工作室建筑创作（2004—2014）》一书的成果及出版，是近十年来我们设计团队共同协作的结果。

首先，感谢教给我们知识的老师们，是你们的辛勤培育，才让我有机会在建筑设计领域取得一点成绩；其次，感谢信任我给我创作机会的领导和业主们，是你们的支持和协作，使得设计作品走向成功；再次，感谢与我共同创作的同事们、我的研究生们及同学们，有了你们卓绝的团队精神和创新智慧，才使得我们的作品上升到一定的高度。当然，我要特别感谢导师齐康院士对我的时时教导、关心和鼓励！学生终生难忘，深表敬意。

感谢身边刚毕业的三位研究生徐崭青、季群珊、王晓帆在资料整理过程中付出的智慧和汗水，尤其感谢在读研二的学生邓奥博、周璐瑶、方亮等三位研究生在图文整理、外文翻译、编排工作中的出色表现。感谢 QI SHANSHAN 老师出色的英文校核工作！

感谢东南大学出版社戴丽副社长的关心和支持。感谢父母和亲人的养育之恩，特别是我夫人对我工作的支持。

最后，借此向帮助、关心和支持过我的各位长辈、同事、朋友致敬！

高裕江

2016 年 6 月 18 日

Postscript

This book *Appropriate Construction* is the collaborative accomplishment of my team in the past 10 years.Many people have made invaluable contributions, both directly and indirectly to my work.

First of all, I would like to express my warmest gratitude to my teachers. It's the knowledge they've passed on me that gives me the ability to score some achievements in architectural design area. I'm also grateful to my leaders and clients for their support in work, which leads the result to success. High regards shall be paid to my colleagues, graduate students and classmates. It's their excellent cooperation and innovation mind that pushes our work to a higher level. Special thanks should go to my most respected tutor Academician QI Kang for his consistent guidance, care and encouragement. I benefit deeply from it.

Besides, I also express my gratitude to my graduate students. Thank Xu Zhanqing, Ji Qunshan, Wang Xiaofan for their efforts in file organization for this book. Special gratitude should be paid to the graduate candidates Deng Aobo, Zhou Luyao and Fang Liang for their outstanding work in file organization, English translation and typesetting. At the same time, thank Professor Qi Shanshan for her excellent English proofreading.

Also, I would like to express my heartfelt gratitude to the care and support from Dai Li, proprieter of Southeast University Press. I own my sincere gratitude to my beloved family for their company and care, especially my wife for her support to my work.

My heart swells with gratitude to all the people who have ever helped me.

Gao Yujiang

18th, June, 2016

图书在版编目（CIP）数据

中庸建构：浙江大学建筑系高裕江工作室建筑创作：2004—2014 / 高裕江著. -- 南京 ：东南大学出版社，2016.6
ISBN 978-7-5641-6565-9

Ⅰ. ①中… Ⅱ. ①高… Ⅲ. ①建筑设计－作品集－中国－2004－2014 Ⅳ. ①TU206

中国版本图书馆CIP数据核字(2016)第157652号

书　　名　中庸建构——浙江大学建筑系高裕江工作室建筑创作（2004—2014）
著　　者　高裕江
责任编辑　戴　丽
文字编辑　贺玮玮
责任印制　张文礼
出版发行　东南大学出版社
社　　址　南京市四牌楼2号（邮编：210096）
出 版 人　江建中
网　　址　http://www.seupress.com
印　　刷　上海雅昌艺术印刷有限公司
开　　本　787mm×1092mm　1/12
印　　张　17 2/3
字　　数　360千字
版　　次　2016年6月第1版
印　　次　2016年6月第1次印刷
书　　号　ISBN 978-7-5641-6565-9
定　　价　198.00元
经　　销　全国各地新华书店